D1375504

Hadlow
c o l l e g e

Date of Return	Date of Return	Date of Return
O/S 1 JUN 11		
-9 JAN 2012		
11 JAN 2012		
2 2 JAN 2015		
-9 FEB 2015		
1 1 FEB 2015		
25 /1/16		
2 2 MAR 2016		
29 OCT 2019		

Please note that fines will be cha

OTTERS
ecology
and conservation

C. F. Mason

and

S. M. Macdonald

Department of Biology, University of Essex

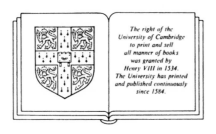

The right of the
University of Cambridge
to print and sell
all manner of books
was granted by
Henry VIII in 1534.
The University has printed
and published continuously
since 1584.

CAMBRIDGE UNIVERSITY PRESS

CAMBRIDGE

LONDON   NEW YORK   NEW ROCHELLE

MELBOURNE   SYDNEY

CAMBRIDGE UNIVERSITY PRESS
Cambridge, New York, Melbourne, Madrid, Cape Town, Singapore, São Paulo, Delhi

Cambridge University Press
The Edinburgh Building, Cambridge CB2 8RU, UK

Published in the United States of America by Cambridge University Press, New York

www.cambridge.org
Information on this title: www.cambridge.org/9780521307161

First published 1986
This digitally printed version 2008

A catalogue record for this publication is available from the British Library

Library of Congress Cataloguing in Publication data
Mason, C. F.
Otters: ecology and conservation.
Bibliography: p.
Includes index.
1. Otters–Ecology.   2. Wildlife conservation.
I. Macdonald, S. M.   II. Title.
QL737.C25M318   1986   599.74'447   85-11711

ISBN 978-0-521-30716-1 hardback
ISBN 978-0-521-10134-9 paperback

Contents

Acknowledgements

We would like to thank Vivien Amos for typing the text and Linnet Barnes for developing and printing a large number of photographs. Photographs of otters were kindly provided by Nicole Duplaix, James Estes, Dennis Hada, Helmut Pechlaner and David Rowe-Rowe. Our own photographs of otters were all taken in captivity, mainly at the Otter Trust. The illustrations concluding each chapter were drawn by Moira Williams. Many people provided us with unpublished information and comments for which we are very grateful; these people are named at the appropriate places in the text.

1 INTRODUCTION

Yet, if you enter the woods
Of a summer evening late,
When the night-air cools on the
trout-ring'd pools
Where the otter whistles his mate
(They fear not men in the woods
Because they see so few).

RUDYARD KIPLING,
The Way through the Woods

In the seventeenth century, Izaac Walton fished the Midland and Hampshire rivers and streams for half a lifetime and was so absorbed by the otter and its ways that he could not shelve a hunting story beyond the first few pages of *The Compleat Angler*. Walton was both humorist and teacher and he asks his Otter Huntsman whether he hunts beast or fish. 'Sir', said the Huntsman, 'it is not in my power to resolve you, I leave it...to the College of Carthusians who have made vows never to eat flesh'. But despite the gentle humour and perennial charm of one of the best loved books in our language, it does not even occur to Walton to hide his true feelings when, at a day's hunt, he adds, 'God keep you all, gentlemen, and send you meet this day with another Bitch-otter, and kill her merrily, and all her young ones too' (Walton, 1653).

Over the centuries, otters have been killed as vermin and bounties paid for them, as for example under the 1566 'Acte for the preservation of Grayne'. Records of bounties were kept by church wardens. Otter hunting with dogs was recorded as early as the reign of King John and even today some will still insist that this is an effective way of controlling 'vermin'. However, despite such persecution, the otter remained widespread in Britain well into the present century.

During the latter half of this century there has been a huge increase in interest in wildlife conservation and the otter is now viewed in a benevolent light. Yet, despite changing attitudes, otter numbers declined severely in Britain over the last 30 years and, in some European countries, the species is now close to extinction. Unwittingly, man has achieved in three decades what centuries of direct persecution could not. The activities that have caused this decline include the widespread use of agricultural chemicals and the discharge of industrial wastes, which can ultimately poison our waterways. Wetlands have been drained to increase cereal production and rivers and streams that were formerly rich with bankside vegetation have been scraped into lifeless drains. Modern methods of industry and agriculture have certainly proved more lethal to the otter and to wildlife generally than the deliberate persecution of earlier times.

The distribution and status of the otter in Britain was long a matter of guesswork and even in 1969 the Mammal Society, by a curious twist of irony, could only obtain the required information from the number of animals found or killed by the otter hunts. Results of field surveys for otter distribution throughout much of the British Isles have now been published and these should prove invaluable for comparison with similar records in future years. Only by patient field work can we begin to understand the otter and its requirements and to unravel the variety of human activities that still threaten depleted populations. During the last 20 years some such studies have been attempted and these will form the substance of this book.

Much of our knowledge of the ecology and behaviour of the otter was little more than hearsay before Sam Erlinge's pioneering studies on Swedish populations in the 1960s. Our understanding is still sketchy and modern research techniques, such as radio-telemetry, have been applied only in the last few years. The otter is notoriously difficult to study. It is largely nocturnal over most of its range. An individual's home-range may include 40 km of river and the males at least appear to live a peripatetic existence within this range. They are mainly silent. They have proved extremely difficult to catch for study. But from what we do know, it seems that their ecology and behaviour show marked variations between different populations, so that studies over a range of habitats will be required before scientific knowledge can be applied as the basis of a management strategy.

Unfortunately, because the otter has declined over much of its range, it is in urgent need of measures to conserve it. Management strategies must be based on detailed knowledge of the species, but if we wait for the completion of fundamental research, there may, in many areas, be no otters left to benefit. Conservation efforts will have to be based on

Fig. 1.1. A car sticker, used to promote the 'Oasi di Protezione Serre-Persano' of the Italian World Wildlife Fund. This wetland reserve, near Salerno, is one of the last strongholds of the otter within Italy.

Fig. 1.2. A logo used by the Council of Europe to promote its extensive activities in the field of nature conservation.

what we know now, though strategies must always be modified in the light of new findings.

But why conserve the otter? Myers (1983) has detailed how many wild species are valuable or potentially valuable in agriculture, medicine or industry, but these uses will hardly apply to the otter unless one considers the production of fur coats as the legitimate end-product of an animal. However, we can see the otter as an indicator of the health of the wetland environment. The species has proved very sensitive to both pollution and to the destruction of riverside habitat. A river supporting a viable population of otters is therefore likely to be in an ecologically healthy condition.

We must also remember the ethical and aesthetic arguments for conservation, but along with this there is the problem that individuals have different ethical standpoints and aesthetic appreciation, and these values cannot be financially quantified. The ethical argument, simply stated, is that all forms of life on earth have a right to exist and that humanity has no right to exterminate a species (Myers, 1979). The aesthetic argument centres around the pleasure of seeing wild creatures, or of knowing that they are sharing the earth with us. Such a view may tend towards the sentimental, particularly with a creature such as the otter. For example, the otter in Kenneth Grahame's *Wind in the Willows* is portrayed as a jocular, playful fellow, but with a mysterious, unpredictable personality. Many a hardened anti-sentimentalist, however, on reading Henry Williamson's gripping depiction of the life of an otter, has been temporarily distressed by the brutal end of Tarka.

A recent M.O.R.I. opinion poll, conducted in January 1983, provided strong evidence of the concern of the general British public to the environment and conservation. When asked which qualities made a valuable contribution to the overall quality of life, 52% of respondents answered 'an attractive countryside', 51% 'an unpolluted atmosphere', 38% 'unpolluted lakes and rivers' and 37% 'wildlife'. Concern was uniform over all sectors of the public, regardless of age, socio-economic class or political affiliation.

The otter appears to have caught the imagination of people with a particular commitment to conservation and has become a symbol of the problems facing threatened wildlife (Macdonald and Duplaix, 1983). It is an emblem used to promote conservation at all levels, from the local (Fig. 1.1) to the international (Fig. 1.2).

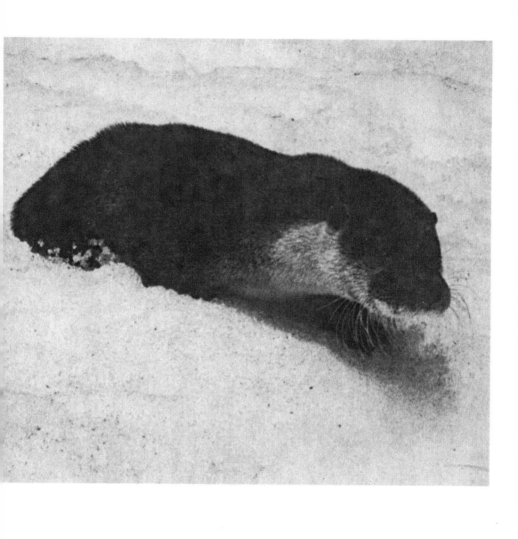

2 *LUTRA LUTRA*

Water-wayfarers

D. H. LAWRENCE,
Fish

Otters make up the sub-family Lutrinae of the family Mustelidae, which, with some 67 species (Ewer, 1973), is one of the largest families of the order Carnivora. The Mustelidae also includes the sub-families Mustelinae (weasels, martens, polecats), Melinae (badgers), Mellivorinae (honey badger) and Mephitinae (skunks). They range in size from the weasel (*Mustela nivalis*), averaging about 100 g, to the sea otter (*Enhydra lutris*), at 30 kg some 300 times heavier. The family is largely carnivorous, provided the term is applied broadly – the badger (*Meles meles*), for example, is an earthworm specialist (Kruuk and Parish, 1981). The Lutrinae are all more or less amphibious.

The European otter has a broad range (Fig. 2.1), from Ireland in the west to Japan in the east, and the arctic north to the semi-desert of North Africa. Within this range, 10 subspecies are recognized, with four others of doubtful validity (Table 2.1).

The body of the otter is elongate and the head flattened, with small ears. The broad muzzle is surrounded by long, stiff whiskers (see frontispiece to this chapter), which may assist the otter in locating prey in murky waters (p. 13). The tail is long, slightly flattened, and tapering, with a thick base, which functions as a rudder during swimming. The legs are short and the five-toed feet are webbed (Fig. 2.2). The muscles in the feet are modified to regulate the tension in the webs (Burton, 1979).

The otter's fur is thick and waterproof, with two types of hair. There is a dense underfur, with hairs 10 to 15 mm in length, which traps an insulating layer of air and remains dry while the otter is swimming. The longer, overlying hairs are 25 mm long and waterproof. The pelage is a uniform brown colour above, the underparts are lighter. A more detailed description is given by Harris (1968).

7

Table 2.1. *The subspecies of* Lutra lutra

Name	Distribution
lutra	Palearctic
aurobrunnea	Nepal
barang	Thailand, Vietnam, Malaysia, Sumatra, Java
chinensis	China
kutab	Kashmir
meridionalis	Iran, southern U.S.S.R.
monticola	Nepal, Sikkim, Assam
nair	Sri Lanka, southern India
seistanica	Afghanistan, U.S.S.R.-Pamir
whiteleyi	Japan

Note: Four other subspecies (*angustifrons, roensis, splendida, stejnegeri*) are of doubtful validity.

Fig. 2.1. The approximate range of the European otter, *Lutra lutra*. The northern limit of distribution is at the Arctic circle, but it extends further north in the north-west of its range and in parts of Asian U.S.S.R. The distribution in Asian U.S.S.R. is not adequately recorded. Several subspecies are recognized within this range (see Table 2.1).

The weight distributions of a sample of males and females are shown in Fig. 2.3 and the relationship between length and weight in Fig. 2.4. The males achieve a larger overall size, being some 28% heavier, on average, than the females. This sexual dimorphism is less than that of weasels (*Mustela nivalis*) and stoats (*Mustela erminea*), where males are often twice as heavy as females. It has been suggested that small females are favoured, because they need less energy for maintenance and can divert more energy into reproduction, while large size in males is due to sexual selection, giving enhanced dominance and mobility (Erlinge, 1979; Moors, 1980). The sexual difference is greater with small mustelids, which have relatively brief lives. Otters are larger, long-lived and have a smaller litter size, and dimorphism is less strongly developed.

Fig. 2.2. A 5-week old otter, showing the five digits and webbing between the toes (photo: E. Hofer/Alpenzoo, Innsbruck).

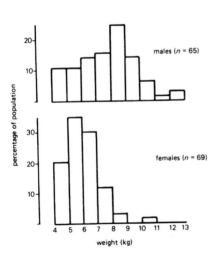

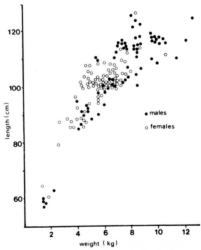

Fig. 2.3. Weights of adult otters, over 4 kg. Mean weights are, for males 7·9 kg, and, for females 6·0 kg. *n* = number of otters (Data from Jensen, 1964; Fairley, 1972; Reuther, 1980*a*; Twelves, 1982.)

Fig. 2.4. Relationship between length (nose to tail measurement) and weight of male (*n* = 77) and female (*n* = 78) otters. There was no significant difference between the length:weight relationship of the two sexes. Data sources as for Fig. 2.3.

Fig. 2.5. Some typical otter habitats: (a) an upland river (Xanthi River, Greece); (b) a small eutrophic lake in Shropshire, with dense vegetation on islands and banks; (c) marine loch in western Scotland.

Otters occasionally grow much larger. Harris (1968) lists females up to 12·6 kg and a number of males up to 16 kg. Records of even larger males must be treated with caution (from the relationship in Fig. 2.4, the 27 kg male mentioned by Harris would be 220 cm, over 7 ft long!).

On land, the otter moves with a bounding gait, rather heavy-footed, but over short distances is as fast as a man (Corbet and Southern, 1977). As an adaptation to life in water, the lungs have a large volume, the right lung having four lobes and the left two. During diving, the nostrils and ears are closed and the heart-rate slows. Dives are usually short, generally less than a minute, but with a top speed of 10 to 12 km/h, the otter may travel up to 400 m without surfacing (Corbet and Southern, 1977).

Otters live along rivers, streams, lakes and marshes. The race *kutab* has been recorded at an altitude of 4120 m in Tibet. Conversely, in some parts of their range, otters spend most of their time on sea coasts. The coasts of western Ireland, western and northern Scotland, and Norway seem especially suitable. Intertidal salt marshes are also visited for feeding. Some typical habitats are shown in Fig. 2.5. For lying up and breeding, the otters require ample cover, but many forms of shelter may be used. On coasts, dens (holts) are found among rock falls or in caves, and in peat banks holts can be extensive and resemble badger setts. Broken rock may be used on river banks. Otters also find shelter within the root systems of mature bankside trees. Piles of flood debris on river banks, field drains and broken masonry are also used as resting sites. In quiet places, otters sleep above ground in dense riverside scrub or in reedbeds, where they construct 'couches' or nests of reeds and grass (Hewson, 1969). Dense growths of bramble (*Rubus*) are particularly important. The range of habitats used by otters will be described in more detail in later chapters.

Foraging and diet

The hunting behaviour of the otter has been studied in captive animals by Erlinge (1968*a*), and Wayre (1979*a*) donned a wet suit to follow his otters as they hunted in a local river. Otters are active hunters, making long underwater searches and exploring along the water's edge, amongst weed beds and tree roots. They become more active when prey is spotted and pursuits may last for two to three minutes, the otter remaining submerged for periods of up to 30 s (Fig. 2.6). It was observed that otters often swam beneath a shoal of fish and the slowest member of the fleeing

Fig. 2.6. Otter pursuing roach beneath the water (photographed in a research arena of Aktion Fischotterschutz, Oderhaus, West Germany).

shoal was pursued. Fish in the wild would often seek cover in weed-beds or behind boulders, where they were easily caught (Wayre, 1979a). Otters usually seize their prey with the jaws, but occasionally a fish is caught in the paws. Otters will also sometimes use their muzzles to turn over stones to seek out crayfish.

Foraging has been studied in coastal otters by Watson (1978) in Shetland. He found that otters gradually worked their way around the coast, with periods of intensive feeding in restricted areas. Bouts of foraging would last up to an hour. By contrast, on the west coast of Scotland, Kruuk and Hewson (1978) found that otters would often swim from their holts for distances of up to 700 m, where they might then forage for up to two hours in small areas of no more than 50 m × 100 m. On a freshwater loch in north-east Scotland, Jenkins (1980) recorded the longest individual period of hunting to be 44 minutes.

Dives have been timed by Hewson (1973) and Jenkins (1980) in freshwater lochs, while Watson (1978) and Kruuk and Hewson (1978) have made similar observations on the coast. The longest dive recorded in these studies was of 45 s. Kruuk and Hewson found that successful dives were shorter on average than unsuccessful dives (15·9 s versus 24·8 s), whereas no such difference was recorded by Jenkins (1980). However, Jenkins did find that dives in summer were longer, on average, than those in winter, which he put down to the summer growth of aquatic weeds making foraging more difficult. The average period of time spent foraging on the sea bottom during a dive sequence was 49% (Kruuk and Hewson, 1978), while an average success rate during foraging was one food item every 29 minutes (Watson, 1978), based on some 90 hours of observation of a female and her cub. In a loch, Jenkins (1980) found that, on average, otters were successful after a period of diving of about five minutes. The length of dives and the hunting success will be influenced by a number of factors, including the depth and clarity of the water, the amount of weed or other potential obstructions, and the individual skills of the otter. The substrate is possibly also important. Otters in Shetland forage mainly among beds of seaweed, avoiding areas of the seabed consisting mainly of bare sand (Watson, 1978).

The time spent on the surface between dives is variable. Sometimes an otter will emerge vertically, loop and plunge head-first back into the water (Watson, 1978); probably the animal had located prey, but was too short of breath to press home the attack on the first dive.

From his observations on captive animals, Erlinge (1968a) considered that otters hunted primarily by sight. However, the vibrissae are probably important in assisting the otter in hunting, at least in murky waters. Green (1977) cut the vibrissae off a captive otter and found that its hunting success was significantly reduced in dark water compared with in clear water.

Hungry otters in the small enclosures of a zoo would attack whatever prey they caught sight of, but when moderately full they would sometimes play with food and occasionally kill fish, but not eat them (Erlinge, 1968a). However, otters were never observed to kill fish and then leave them uneaten in the field (Erlinge, 1967a).

Erlinge (1968a) made observations on feeding preferences of his captive otters, confirming his conclusions from field work that the choice of prey is determined largely by availability and vulnerability. Roach (*Rutilus rutilus*), perch (*Perca fluviatile*) and pike (*Esox lucius*) were caught early, but rainbow trout (*Salmo gairdneri*) were more difficult to catch. Nevertheless, moving fish were preferred to the less mobile frogs and crayfish (*Austropotamobius pallipes*). In fish of mixed sizes, specimens of intermediate size (15–17 cm) were taken before smaller ones. However, Wise, Linn and Kennedy (1981) compared the size of food eaten with estimates of prey size, made by a variety of fishing techniques, and concluded that no size selection was occurring.

Erlinge (1968a) observed that captive otters sometimes ate small fish, less than 10 cm long, in the water, but larger fish were always consumed ashore. The fish is always eaten head first, thus rapidly killing the prey, but the heads of large specimens may sometimes be discarded. Frogs were also eaten head first, but crayfish were crushed, the tail or thorax then being consumed. Watson (1978) noted that crabs were always brought ashore. They were turned on their backs and held down with the forepaws, while the chelae were torn off. The underside of the crab was pulled upwards until it separated from the carapace. The underparts of the crab were then eaten, but the legs were frequently left. The flesh on the underside of the carapace was finally scraped off, apparently by the incisors.

Food is chewed from the side of the mouth, with six to ten chews on each side, the otter changing sides with a quick jerk of its head (Erlinge, 1968a). Both in captivity and in the wild there are few remains from meals. Watson (1978) found the lumpsucker (*Cyclopterus lumpus*) to be an exception, this fish having a thick layer of cartilaginous tissue beneath the skin. Those remains that are left are scavenged by sea eagles (*Haliaeetus albicilla*) in Norway, where this raptor is still widespread (Love, 1983). In the past, Orkney islanders would search the haunts of otters in the early morning for the bodies of conger eels, of which the otter would eat only a small portion (Yarrell, 1841).

The diet of otters has been quite well studied; indeed diet is the easiest aspect of otter biology to study. Most work has been done on the analysis of *spraints* (the term specific to otter faeces or scats), though some biologists have made use of the guts of otters killed either by hunters or accidentally. There are some drawbacks in assessing diet from spraints. The spraint contains those parts of the food intake that cannot be digested (scales, hair, feathers, bones etc.). Food which consists mainly of soft parts will be under-represented in the sample. However, large numbers of samples can be collected without killing otters.

There are several ways of expressing the data obtained from spraint analysis. Most often, the percentage frequency or relative frequency is determined. The percentage frequency records the percentage of spraints containing a particular prey item. With relative frequency, the number of occurrences of an item is expressed as a percentage of the total number of occurrences of all items in the sample, the sum of the frequencies being 100%. In these methods, the occurrence of a single item in a particular spraint, e.g. a single trout vertebra will be weighted the same as the presence of a number of those items in a spraint, e.g. several trout vertebrae and trout scales. This may lead to minor items being over-represented and major items underestimated. To get over this problem, the importance of a particular item in a spraint can be scored visually, for example on a scale of one to ten, so that the total score for each scat is 10. The score for each item is multiplied by the dry weight of the spraint, and the values for each item in each sample are summed and expressed as the bulk percentage (Wise et al., 1981).

Erlinge (1968*a*) fed mixtures of food to captive otters and analysed the resultant spraints. He concluded that spraint analysis, calculated by relative frequency, gives a reasonably true picture of the relative importance of the different food categories. Webb (1976) and Camby, Le Gall and Maizeret (1984) describe the method of analysing otter spraints and provide a preliminary key to the identification of freshwater fish scales, vertebrae and teeth. Watson (1978) provides a similar key for intertidal fish while Day (1966) gives information on the identification of fur and feathers. Wise (1980) describes a method of relating the size of individual vertebrae in the spraints to the lengths of fish eaten by the otter.

The results of a number of dietary studies are given in Table 2.2, and Fig. 2.7 illustrates the diet from four contrasting habitats. It is obvious that otters are largely piscivorous, the majority of studies recording more than 80% of the diet as fish in northern Europe. Authors reporting lower proportions of fish have used the percentage frequency, rather than bulk percentage method and have included all invertebrates found, though many of these are small items derived from the stomachs of fishes on which the otter was feeding; they are insignificant in terms of the weight of food eaten. Plant material has been omitted from consideration because it is likely to have been ingested entirely by accident.

Of the fish taken, eels (*Anguilla anguilla*) appear to be important in all habitats (Fig. 2.7). Cyprinid fish feature heavily in eutrophic lakes and rivers, with salmonid fish being important in oligotrophic rivers and lakes. In lowland rivers and dykes, three-spined sticklebacks (*Gasterosteus aculeatus*) are often abundant and, although usually no more than 5 cm long, they may be taken in large numbers by otters.

Table 2.2 demonstrates the overall proportions of fish in the diet of otters, but there are seasonal differences, which may be related to both availability and to the preferences of otters. Figure 2.8 illustrates the seasonal variation in the proportions of the major fish species in the diet of otters feeding in a eutrophic lake and an oligotrophic river. In both habitats, eels were taken least in winter. In north-east Scotland, there was a decline in the numbers of eels taken in the river during the winter, with an increase in trout, but eels were the most important food throughout the year in a loch (Jenkins and Harper, 1980). In Slapton Ley, roach became more important in winter, while trout (*Salmo trutta*) were taken in greater proportion in the river. Eels become torpid during the winter and may be difficult to find, otters then turning to other prey. Most roach and trout taken in these Devon habitats were less than 12 cm long, whereas most eels were 25 to 35 cm long (Wise *et al.*, 1981).

Table 2.2. Diet of otters in freshwater habitats

Locality	Habitat	Method	Sample size	Total fish	Cyprinidae	Percidae	Esox lucius	Gasterosteus aculeatus	Salmonidae	Anguilla anguilla	Other fish	Amphibians	Reptiles	Mammals	Birds	Invertebrates	Reference
Great Britain																	
Slapton Ley, Devon	Eutrophic lake	S/B	1547	92.8	45.4	10.6	9.2	0	0	26.6	1.0	0.5	0	1.2	4.6	1.0	1
Slapton Ley, Devon	Eutrophic lake	S/A	607	91.4	32.1	28.2	6.3	1.5	0.2	23.2	0	0.5	0	1.0	7.1	0	2
R. Dart catchment, Devon	Oligotrophic river	S/B	675	81.7	0	0	0	0	59.1	16.2	6.4	5.4	0	6.8	1.6	4.4	1
R. Teign, Devon	Oligotrophic river	S/A	353	91.8	0	0	0	0	60.1	29.5	2.3	0.6	0	4.2	1.1	0	2
Various	Various	G.S/A	110	68.7	11.2	3.1	0.4	7.1	21.1	18.4	7.4	7.8	0	9.6	11.0	0	3
Somerset Levels	Eutrophic rivers	S/A	858	66.2	9.4	2.2	1.4	25.7	0.9	25.0	9.9	7.8	0	0.4	4.7	20.1	4
Somerset Levels	Eutrophic rivers	S/A	120	80.0	12.0	3.2	3.2	8.0	0	54.4	0	2.0	0	4.0	5.0	9.0	5
Bosherston Lake, Dyfed	Eutrophic lake	S/A	258	99.4	38.0	0.7	2.2	7.7	0.2	50.6	0	0	0	0	0.7	0	6
Blakeney, Norfolk	Eutrophic river/ brackish marsh	S/A	2260	83.0	12.4	1.7	1.4	33.7	1.7	26.5	5.6	1.0	0	1.7	10.3	4.1	7,8
Ireland																	
Agivey River, Derry	Mesotrophic river	S/A	84	88.0	5.2	11.5	1.0	0	63.0	7.3	0	7.3	0	0	5.2	0	9
Galway, Eire	Various	G/A	29	67.2	3.3	16.4	8.2	1.6	6.6	31.1	0	13.1	0	0	1.6	18.0	10
Lough Feeagh, Mayo	Oligotrophic lake	S/A	266	84.3	0	0	0	5.1	39.8	33.3	6.1	3.7	0	0	1.0	11.0	11
Lough Furnace, Mayo	Brackish lake	S/A	440	85.7	0	0	0	13.9	13.5	43.2	15.1	1.0	0	0	1.0	12.4	11
Sweden																	
Various	Eutrophic lakes and streams	S/A	14615	66.9	34.2	16.5	9.3	1.0	0	1.7	4.1	8.0	0	0.6	9.2	15.3	12
R. Svartån, Östergötland	Oligotrophic river	S/A	350	93.2	23.1	8.6	14.1	0	5.3	2.8	39.5	1.6	0	0.1	3.9	0.9	13

Table 2.2. (contd.)

			n															
Denmark Various	Various	G/A	61	90.2	18.6	8.9	5.3	10.6	9.7	23.0	8.8	9.7	0	0	2.6	3.5		14
Spain Galicia, N.W. Spain	Rivers	S/B	260	75.5	22.3	0	0	0.1	42.1	7.3	5.1	7.3	2.3	10.3	4.4	0.2	15
Portugal Mondego River catchment	Eutrophic river	S/A	63	44.9	—	—	—	—	—	—	—	31.3	5.1	0	1.7	16.9	16
Greece Various	Various	S/A	80	54.9	—	—	—	—	—	—	—	20.9	0	7.7	6.6	9.9	17

Note: Information is presented as percentage of total items in sample. Methods: S = sprint analysis, G = gut analysis, A = relative frequency, B = bulk percentage (see text). Some data have been recalculated from the original sources. Vegetation has been omitted, on the assumption that it was taken incidentally. For the same reason, some authors do not record invertebrates other than crayfish. References are as follows: 1, Wise *et al.* (1981); 2, Chanin (1981); 3, Stephens (1957); 4, Webb (1975); 5, Jarman (1979); 6, Henshilwood (1981); 7, Weir and Bannister (1973); 8, Weir and Bannister (1977); 9, Fairley and Wilson (1972); 10, Fairley (1972); 11, Gormally and Fairley (1982); 12, Erlinge (1967a); 13, Erlinge (1969); 14, Erlinge and Jensen (1981); 15, Callejo Rey *et al.* (1979); 16, Simões Graça and Ferrand de Almeida (1983); 17, Macdonald and Mason (1982a).

Some important studies (e.g. Jenkins *et al.* 1979; Jenkins and Harper 1980) have had to be omitted from the table because they could not be cast into the summary format.

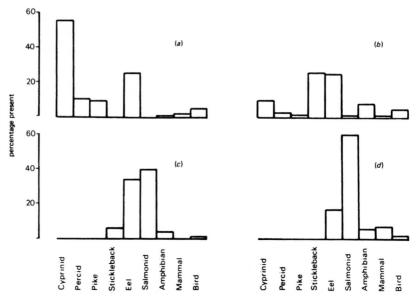

Fig. 2.7. Frequency of the major components of otter diets in four habitats. (a) an eutrophic lake (Slapton Ley); (b) an eutrophic river system (Somerset Levels); (c) an oligotrophic lake (Lough Feeagh) and (d) an oligotrophic river (River Dart). Data from Webb (1975), Wise *et al.* (1981) and Gormally and Fairley (1982). See Table 2.2 for details.

Similarly, most salmonids eaten in north-east Scotland were less than 13 cm long, whereas most eels were 22 to 42 cm in length (Jenkins, Walker and McCowan, 1979). In the River Dart, eels formed a greater proportion of the diet of otters relative to their abundance in the river, suggesting selectivity. In warm weather, fish such as roach and trout move more quickly and may be more difficult to catch.

Migrating salmonids spawn in the headwaters of river systems and otters appear to follow them up tributaries. Most pike (*Esox lucius*) were taken in Slapton Ley in spring, when they spawn. Trout and pike may be easier to catch after spawning, when they are weaker.

Although amphibians are frequently eaten by otters in southern Europe (Table 2.2), they comprise only a small proportion of the overall diet in the north. Jenkins and Harper (1980) found most in spring and otters are known to travel to upland marshes, ponds and headwaters to feast on spawning amphibians (Green and Green, 1980). Fairley (1984) suggested that frogs formed the bulk of the otters' diet at this time in his study area in western Ireland. Otters also, however, forage for hibernating frogs (*Rana temporaria*) during the winter. We have frequently

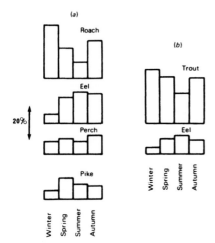

Fig. 2.8. Seasonal variation in bulk percentage of the major prey animals in the diet of otters at (a) Slapton Ley lake and (b) River Dart catchment, based on spraint collections made over two years. (Adapted from Wise et al. 1981.)

found the spawn of frogs, which appears not to be eaten, smeared with blood and bits of skin and marked with spraints, by riverside ponds during January. In southern Europe, where amphibians are numerous and several species are permanently resident by watersides, they form a significant part of the diet. For example amphibians made up 31% of the diet in Portugal (Simões Graça and Ferrand de Almeida, 1983) and up to 24% of the diet in Spain (Callejo Rey and Delibes, 1985), where they were taken in greater proportion in autumn and winter.

Reptiles appear never to be taken in northern Europe, but they form a significant component of the diet in southern Europe (Table 2.2). We found evidence of snake in 36% of spraints collected in Portugal in the summer (Macdonald and Mason, 1982b). They were taken most frequently in spring and summer in Spain (Callejo Rey et al., 1979). In Greece, we found few reptile remains in spraints collected in April, whereas in July some 30% of spraints contained snake (unpublished data). Terrapins are also sometimes taken by otters (Macdonald and Mason, 1982b).

Birds make up only a small proportion of the diet, though more may be taken at eutrophic lakes, where waterfowl congregate. More birds are taken in the summer. Eggs are not eaten (Erlinge, 1968a; Wise et al., 1981). Mammals are also taken in relatively small numbers, though a wide variety of species, up to the size of a rabbit (Oryctolagus cuniculus) may be eaten. Most birds and mammals must be caught as they come down to the waterside to drink and this may explain the remains of species such as jay (Garrulus glandarius) and grey squirrel (Sciurus carolinensis) in spraints (Callejo Rey et al., 1979; Wise et al., 1981).

Many of the small invertebrates (amphipods, molluscs, etc.) found in spraints are probably taken incidentally, being in the guts of otter prey. Earthworms have been recorded in otter stomachs by Fairley (1972), and Lydekker (1896). Earthworms form a substantial part of the diet of other carnivores, such as badger (*Meles meles*) and fox (*Vulpes vulpes*), but earthworm remains (mainly chaetae) in faeces are very small and may be overlooked. We have seen captive smooth-coated otter (*Lutra perspicillata*) digging for and consuming earthworms. We have also found the remains of large bush crickets (Tettigoniidae), almost certainly taken deliberately, in spraints of otters in Greece. Crayfish and crabs are actively hunted by otters, though they mostly form a small proportion of the diet in freshwater habitats. Erlinge (1968a), in studies on captive otters, found that fish were preferred far more than crayfish. Nevertheless, crayfish formed more than 30% of the diet in some habitats in Sweden and Spain, and more than 70% in a river system in western Ireland (Erlinge, 1967a; Callejo Rey and Delibes, 1985; McFadden and Fairley, 1984). We found that freshwater crabs (*Potamon* spp.) made up some 10% of the diet of Greek otters (Table 2.2). In Sweden, crayfish were taken mainly during the summer, when they are most active.

Do otters eat carrion? Erlinge (1968a) found that dead fish were the least preferred prey of captive otters, only being eaten when hungry. Attempts to get otters to take marked baits, in order to study their movements, have always failed. During severe weather deer hairs have been found in spraints from Scotland (Green and Green, 1980), suggesting that otters will take carrion when normal foods become unavailable.

The diet of coastal otters has received little attention as yet. Elmhirst (1938) made observations on otters eating crabs, but provided no quantitative information. Spraints from Shetland, Loch Broom (north-west Scotland) and Norway have been analysed (Table 2.3). The Shetland site of Watson (1978) was an exposed rocky shore, while Herfst (1984) collected spraints from rocky and non-rocky sites. Both the Loch Broom and Norwegian sites were more sheltered, a mixture of rocky headlands and beaches. The Gadidae were important at all sites, while crabs and squat lobsters were particularly significant in the spring diet in Shetland and in winter in Loch Broom and Norway. Crustaceans were taken in very small amounts in the Outer Hebrides (Jane Twelves, unpublished data). Sea Scorpions (Cottidae) were important in Shetland, but were rare in Norway and not recorded in Loch Broom. Flatfish were very scarce in Norway, but the otters here ate a substantial number of

Table 2.3. *The diet (relative frequency, %) of otters feeding in coastal habitats*

	Shetland				Shetland	Loch Broom	Hitra, Norway			
Months...	Jan.	May–June	July–Aug.	Sept.	June–Aug.	Dec.–Jan	Nov.–Dec.	Jan.–Feb.	Apr.–May	June–July
Sample size	95	112	84	146	121	50	56	26	36	90
Eelpout, *Zoarces viviparus*	6·3	9·1	16·9	13·3	21·2	8·6	2·8	5·1	3·3	6·7
Butterfish, *Pholis gunnellus*	2·2	19·7	12·0	18·0	13·5	11·4	15·5	7·6	26·7	20·7
Yarrell's blenny, *Chirolophis ascanii*	1·3	11·9	16·5	7·7	3·4	10·7	0	0	0	0
Cottidae (sea scorpions)	24·6	8·8	14·6	12·6	9·6	0	0	0	1·7	0·7
Fifteen-spined stickleback, *Spinachia spinachia*	6·3	3·0	0	0·7	1·4	2·1	0	0	0	0
Gadidae (rockling, saithe, cod etc.)	30·4	16·4	15·0	17·8	15·8	21·4	33·8	28·2	20·0	12·5
Flatfish (Heterosomata)	10·8	7·9	9·0	19·3	9·1	7·9	0	0	0	1·5
Eel, *Anguilla anguilla*	0·9	1·5	3·4	0·5	5·0	8·6	2·8	5·1	8·3	25·9
Other fish	10·3	5·5	3·8	5·7	16·0	9·3	37·9	15·3	21·7	13·5
Total fish	*93·1*	*83·8*	*91·2*	*95·6*	*95·0*	*80·0*	*92·8*	*61·3*	*81·7*	*81·5*
Crustacea – Decapoda (crabs squat lobsters)	5·8	16·1	7·5	4·0	3·9	17·9	5·6	35·8	18·3	18·5
Birds	0·9	0	1·5	0·5	1·1	2·1	0·2	0	0	0·2
Mammals	0	0	0	0	0	0	1·4	3·0	0	0

Note: Data for Shetland from Watson (1978) and Herfst (1984), for Loch Broom from Mason and Macdonald (1980) and for Norway from Amy Lightfoot (unpublished data).

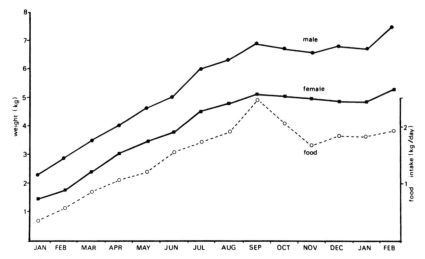

Fig. 2.9. Weight increases of two captive otter cubs, and their daily food intake, over 14 months. The weight loss during January occurred when the cubs were being forced to enter water to feed for the first time, which they did not relish. They also contracted coccidiosis in January, but quickly recovered and began to regain weight. (Adapted from data in Stephens, 1957.)

eels, while herring (*Clupea harengus*) made up 11% of the diet in November and December. Otters may sometimes take seabirds, such as adult puffins (*Fratercula arctica*) from their burrows close to the sea (Harris, 1984). Otters are clearly taking what is available to them within their foraging capabilities. Watson (1978) suggested that the preferred fish were the demersal (bottom-living) species, actively swimming species being taken when demersal fish were scarce. He also noted a difference in the diet between a female and her 10 month old cub, who took many more slow, demersal fish and crabs than the female, presumably due to lack of hunting experience. He noted that the cub hunted for longer than his mother.

How much does an otter eat? Wayre (1979a) fed two captive male otters a mixture of meat-mix and fish over a period of a week and weighed the food that was left. The larger male (12·3 kg) ate 1·5 kg each day, while the smaller animal (10·9 kg) averaged 1·4 kg/day. The captive otters studied by Erlinge (1968a) had satisfied their hunger after having eaten 0·9 to 1·0 kg of food. They may of course have eaten more than one meal each day. Stephens (1957) received two otter cubs at the end of December, when they were about two months old and she recorded their weight each week and the amount of food they ate. The data are summarized in Fig. 2.9. Over the period of 14 months their daily food intake averaged 15% of their total body weight, which compares with the 12·2 to 12·8% of body weight consumed by Wayre's adult males. The high food intake of otters is a result of their high basal metabolic rate, which is some 20% higher than would be expected from the mammalian standard curve relating basal metabolic rate to weight (Iverson, 1972).

With such a high food requirement, there must be some streams, especially oligotrophic waters at high altitudes, unable to support otters. If we assume an average density of otters of one animal per 10 km, living in a stream of average width 10 m and eating 365 kg of food each year, 80% of which is fish, then the animal will require 2·92 g/m² per year of fish *production* to survive, without depleting the standing crop of fish. In the oligotrophic headwaters of the Pennines, trout production averages only 2·5 to 3·5 g/m² per year (Crisp, Mann and McCormack, 1975), while bullhead (*Cottus gobio*) production is only 0·5 g/m² per year (Crisp *et al.*, 1974). Minnows (*Phoxinus phoxinus*) were also present in these streams but production was not determined. Clearly these high altitude, oligotrophic streams would present limited food supplies to resident otters. In Scotland, Green and Green (1980) noted a decrease in marking intensity by otters with increasing altitude, which they attributed to a decline in food supply. It is also of interest that in the River Wye catchment, in Wales, the River Bidno, a headwaters stream, has a trout production averaging 4·6 g/m² per year (Milner *et al.*, 1978) and does not support resident otters (authors' pers. obs.; Andrews and Crawford, 1984). The River Duhonw, joining the main river in its middle reaches, has resident otters and a trout production of 15·8 g/m² per year on average. The production of five other species of fish in these tributaries was not determined. Fish production in lowland rivers may be greater than 40 g/m² per year (Mann, 1969).

In the past, otters have been ruthlessly killed because of their supposed depredations on fisheries, anglers being especially concerned about the damage to trout and salmon stocks. Was this concern justified? To answer this question adequately would require an enormously expensive and long term piece of research. Not only would the numbers of otters and their diet and food consumption need to be known, but data on the populations and production of fish species through the range of the otter would also be required, information which is itself extremely difficult to collect. But the impact of otters on a fishery cannot be measured merely in terms of the weight of the sport fish eaten. Fish populations exhibit both intra-specific and inter-specific competition (Zaret and Rand, 1971; Kerr and Werner, 1980). Intra-specific competition usually results in a lowered growth rate in fishes, resulting in a stunted population. Otters tend to take smaller fish (< 15 cm), which in theory could, by reducing the population, allow other fish to grow to the prime size required by anglers. Inter-specific competition tends to restrict the feeding niches of species, while the eggs and fry of sport fish may be eaten by other species of fish in the community. Managers of trout waters frequently remove coarse fish. The eel is particularly disliked by anglers and it is said to consume large quantities of trout eggs, though the matter is controversial. Otters catch trout with difficulty, especially large ones, and prefer slower-moving fish whenever available (Erlinge, 1968a). In a mixed fishery, they are likely to catch eels and coarse fish, to the benefit of the trout. In the Outer Hebrides, otters feeding in freshwater lochs took many eels, but no trout (Jane Twelves, unpublished data).

Although adequate data on the impact of otters on fisheries are not available, nor probably likely to be, it seems reasonable to conclude that the otter is at at least neutral, and could very well be beneficial, to fishing interests.

Spacing patterns and social organization

Before describing the spacing patterns of otters, we need to define some of the terms that are used. Many mammals live within a *home-range*, which can be defined as an area that an animal learns thoroughly and habitually patrols. The home-range contains the basic requirements for the maintenance of the individual. The term does not imply exclusive use and an integrated social group may live within strongly overlapping home-ranges. Within the home-range there is a core area or centre of activity where an individual spends a disproportionate amount of its time.

An area occupied by an individual or group more or less exclusively, and which is defended from other individuals or groups by overt aggression or advertisement, is a *territory*.

Within a social group there may be a system of dominance. The simplest form is despotism, where one individual is dominant to all other members of the group, who have no rank. More often there is a dominance hierarchy, where the alpha individual dominates all others, the beta individual dominates all but the alpha, and so on. Some animals are organized into absolute dominance hierarchies, which remain constant under all circumstances, unless rank is changed after interactions between rivals. Other species show relative dominance hierarchies, where a dominant animal will yield to a subordinate under certain circumstances. For example, dominant domestic cats will respect the sleeping places of subordinates (Leyhausen, 1971). Within a population, individuals may exist as residents, temporary residents or transients, these classes occurring in all mustelids so far studied (Powell, 1979). Residents stay in an area for extended periods or for their entire life, temporary residents stay for a short time, while transients pass through an area without establishing a home-range. However, individuals will have to attain resident status before they can successfully rear a family.

A number of methods have been used to determine the spacing patterns and population density of otters. Because otters are largely nocturnal, direct observation provides little information. Watson (1978) studied a number of animals in Shetland that were active during the day and could be individually recognized, giving valuable information. The advent of image intensifiers might make nocturnal observations more practicable.

Until recently, most studies have made use of tracks and signs. Otter tracks are distinctive and, by measuring footprints, individuals can occasionally be distinguished. Animals can be followed for long distances across snow, but without snow the observer has to rely on patches of waterside mud, which is most extensive when water levels are falling. It is possible to place wet mud or sand by centres of activity to record footprints.

Spraints are also highly distinctive and are deposited in prominent places. Tracks and signs of otters and their use in surveys are described in Chapter 3, while the social function of spraints is discussed on p. 31. It is sometimes possible to determine the movements of individual animals, if their spraints contain distinctive items. Erlinge (1967*b*) found that some otters had tapeworm parasites and spraints regularly contained parts of the parasite. For some species, e.g. badger (Kruuk, 1978), coloured baits have been incorporated into the food and the appearance of these in the faeces allows the movements of an animal to be followed. Wild otters appear not to take baits, so that attempts to use this method have so far proved unsuccessful.

Other techniques require the trapping of the otter. Hancock livetraps have been most effective (see Melquist and Hornocker, 1983) and otters are normally anaesthetised with ketamine hydrochloride. Live-trapping otters in Britain has proved extremely difficult and the animals appear to become quickly aware of the presence of a trap.

Radio-telemetry is proving an invaluable technique in studying the movements of mammals, particularly those species active mainly at night, but it has been applied only recently to otters. The animal is caught and a small radio-transmitter is attached. The individual can be located at any time with a directional aerial and with two directional receivers its position can be determined with reasonable accuracy. An individual can be recognized by the characteristics of its transmitter signal and several animals can be monitored at once.

The otter has a rather small head and a thick neck so that collars, the usual device for attaching radio-transmitters, tend to slip off, and they may irritate the neck. There is also the danger that collars and harnesses may snag while the otter is under water. To overcome these problems, Melquist and Hornocker (1983), working with *Lutra canadensis*, developed transmitters which could be implanted into the intraperitoneal cavity. The transmitters appear large (9–11 cm × 3–4 cm, weighing 70–130 g) but otters seemed to suffer no ill effects, one female giving birth and rearing cubs while carrying implants. Melquist and Hornocker studied the movements of 39 otters in Idaho by radio-telemetry.

Under British law it is illegal to make such implants into wild animals, so harnesses are attached instead. The rivets holding the harness together are designed to rust through after about eight weeks, releasing the animal. This reduces the risk of an animal being caught up and drowned, but it means that the change in home-range use of an individual cannot be followed through the seasons. Mitchell-Jones *et al.* (1984) provide details of the method. On some animals, beta-lights have been attached to the harness. These are small glass tubes, which contain a radio-active source, which is of low energy and confined entirely within the tube. The beta-light enables the precise position of an animal tracked by radio to be determined.

Otters can also be injected with a radio-isotope that appears in the spraints; ^{65}Zn has been used and can be detected in spraints for a considerable time. Kruuk, Gorman and Parish (1980), with captive badgers, have shown that the proportion of marked to unmarked faeces gave a good estimate of the numbers present and they have applied the technique to the field. However, the method assumes that faeces from labelled and unlabelled animals are equally likely to be found, but this seems unlikely to hold for otters, for sprainting intensity is probably influenced by social status and sex (p. 31). Nevertheless, the use of ^{65}Zn enables the minimum range used by an otter to be determined.

Erlinge pioneered the study of otters in the field. By following tracks and finding spraints, he obtained data over a total of 385 days in Sweden (Erlinge, 1967*b*). His study areas consisted of lakes, joined by streams. The lakes froze over in winter, so that otters concentrated their feeding along the streams. Females with cubs were the easiest to track and they tended to restrict their activity to definite areas, having limited home-ranges. Erlinge determined the home-range of six family groups using both lake and stream. When the water was open they utilized 2 to 4 km² of lake and 1 to 3 km of stream, with a radius of activity of 2 to 4 km. When the lakes froze, they utilized a range of 3 to 6 km of stream, with a radius of activity of 2 to 3 km. A family group with a home range based entirely on the stream used 10 to 12 km. Dog otters used larger areas. The approximate size of the home-range of eight dogs, measured as the travelling length across the home-range in winter, was 10 to 21 km (mean, 15 km).

Erlinge (1968*b*) observed that intense signal activity, i.e. spraint deposits, occurred at the meeting zones of otters and that signal intensity increased at times of high population density. He considered that adult dog otters maintained territories against other dogs. Females with cubs occupied areas separated from one another and not necessarily contiguous. Dog otters usually covered areas that included one family group and part of another, but they paid little attention to families. A hierarchy appeared to exist, with dominant dogs occupying the most favourable parts of an area, subordinates living in less favourable areas. The smaller family territory gradually extended as the cubs grew. Subordinates occupied temporary territories when family territories were small, but they were forced to live a transient existence when family territories were at their largest. Dominant dogs maintained territories by marking with spraint, and other dogs avoided contact with the dominant. However, intrusion into territories occurred frequently, especially around the periphery and occasionally in the centre. When densities were high territories greatly overlapped.

The data presented by Erlinge (1968*b*) do indicate that dog otters defend territories against other dogs, which he considers primarily to have a sexual significance. The territories of the family groups were chiefly feeding areas, giving access to adequate food through the year.

Preliminary data based on otters injected with radio–isotopes and on radio–telemetry have recently become available. In north-east Scotland,

Jenkins (1980) marked a young male with radio-active zinc chloride. When caught, in early October, the animal was thought to be four to five months old, and still with its mother. It remained close to the area of capture, in a loch, until about eight months old, when it was recorded some 4 km distant, on the adjacent River Dee. It then began to move greater distances and, by May, labelled spraints were separated by a distance of 65 km. It was last recorded when about 12 months old, after having visited sites on the river 68 km apart.

Green, Green and Jefferies (1984) and Green and Green (1985) have radio-tracked two adult female otters, one adult male and two young males in Perthshire, central Scotland. The females were also labelled with radio-active zinc. The study area embraced a largely undisturbed system of oligotrophic rivers and lakes. The first female (weighing 6·25 kg) to be captured ranged over 16 km of waterway, comprising the main river, tributary streams and small lakes, during the 22 days of radio-contact. Monitoring of radio-active spraints gave information for a further $3\frac{1}{2}$ months, during which an additional 2 km of river were added to the range. The second female showed signs of old injuries and weighed only 4·2 kg, low for her size. She utilized 22·4 km of waterway over 36 days of radio-contact and monitoring of radio-active spraints for a further four months added a further 2·4 km of river.

The adult male (weight 9 kg) was radio-tracked for 98 days and used a home-range of 39·1 km of waterway. This animal also used an overland route of 3 km to reach a marsh. His range overlapped with that of a female for 5 km. The two young males had home-ranges of 20 km and 31·6 km of waterway, respectively.

The boundaries of home-ranges tended to coincide with significant features of habitat, such as confluences of streams. The animals did not utilize all of their home-ranges evenly (Fig. 2.10). For example, the first female spent 73% of the 22 days of radio-contact in one lake, which comprised 58% of the home-range, two other lakes forming a secondary centre of activity, the stream joining the two centres being used for travelling. The second female spent 42% of her time in a marsh of 3 km². The male otter spent 60% of the period of radio-contact in a stretch of 10·7 km of the river, which represented 27% of his linear home-range. Within this, there appeared to be two centres of activity, each consisting of about 2 km of main river and 2 km of tributary stream.

The adult male was the only animal that showed an obvious interest in the boundaries of his home-range. He spent more time at the downstream boundary of this range, which was judged to have especially

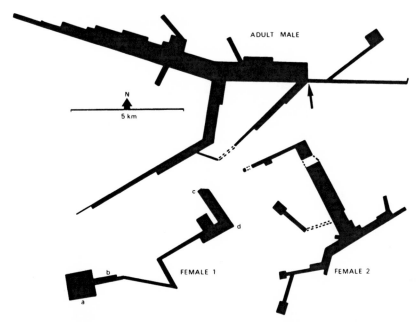

Fig. 2.10. Utilization of home-ranges by three otters in Perthshire, Scotland. Line thickness is proportional to the number of daily journeys made through each area during the period of radio-contact. Arrow indicates the downstream boundary of male's normal home-range, dotted lines represent overland routes, and letters a–d indicate lakes utilized by female 1. (Adapted from Green *et al.*, 1984.)

high otter activity by Green *et al.* (1984). The upper boundary was visited less often and the short duration of the visits suggested that they served a communicatory function, rather than feeding.

Although few otters were radio-tracked, Green *et al.* (1984) made other observations that enabled them to suggest a probable spacing pattern in this population. They considered that adult females had overlapping home-ranges, with a degree of mutual exploitation of resources, which may have varied with the productivity of the habitat. Breeding females may have had temporary dominance over both non-breeding females and adult males. The social system of females in Scotland thus appears to be less rigid than those in Sweden. Green *et al.* believe that the relationship between males in their study area is more rigid than between females, dominant males maintaining relatively exclusive territories, with sub-dominant males utilizing sub-optimal habitat. This resembles the situation described by Sam Erlinge in Sweden. The dominant male exploited the major thoroughfares of the river and had access to several females.

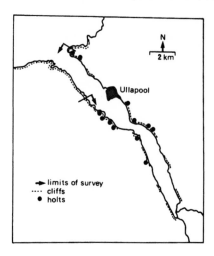

Fig. 2.11. Position of occupied holts along the shore of Loch Broom, western Scotland. (Adapted from Macdonald and Mason, 1980.)

Comparison can be made with results of radio-tracking 39 Canadian otters (*Lutra canadensis*) over a five year period in Idaho (Melquist and Hornocker, 1983). Home-ranges were similar to those of *Lutra lutra* (p. 165), with males holding bigger ranges, up to 63 km. There was extensive overlap between female home-ranges, while adult males also appeared to overlap considerably in range. As with *Lutra lutra*, there were centres of activity within home-ranges.

Coastal otters appear to live at a higher density than otters in freshwater habitats. Watson (1978) made sufficient observations on a female and her cub in Shetland to delineate the home-range as 2·5 km of shoreline. As they fed within 150 m of the shore, the feeding zone encompassed an area of 0·37 km². This can be compared with the 0·62 and 1·36 km² of aquatic habitat for females and 2·76 km² for the male in the Perthshire study of Green *et al.* (1984). Recent radio-tracking of coastal otters in the Outer Hebrides has confirmed that they have much smaller ranges than do river otters (Jane Twelves, unpublished). Evidence from the distribution of holts and high density of sprainting sites (see below) also confirm a high population density of otters. Thus otter holts on the Ardnish peninsula, Inverness-shire were regularly spaced out at an average distance of 1·1 km (Kruuk and Hewson, 1978). The inter-holt distances in suitable habitat, rocky outcrops, at Loch Broom, Easter Ross was 0·94 km (Fig. 2.11), while at Applecross, Wester Ross, holts were 0·75 km apart in suitable habitat (Jenkins and Burrows, 1980). On South Uist, the average distance between holts was 0·59 km (Jane Twelves, unpublished data). Lightfoot (1981) found that occupied

holts were 4 km apart at Hitra on the Norwegian coast. Norwegian holts
were situated near to the exits of streams, where open water for foraging
was maintained throughout the winter. In the Outer Hebrides, holts are
situated where there is good access to the sea at all states of the tide and
adjacent to dense growths of seaweed, containing good prey populations
(Jane Twelves, unpublished).

The significance of spraints

In many vertebrates, including carnivores, scent marking plays an
important role in social organization (Stoddart, 1980; Macdonald, 1980;
Gosling, 1982). Urine, faeces and scent from specialized glands may be
used to delineate territories. The information contained within these
signs may enable individuals to distinguish one another and the signaller
may also indicate its sexual condition or status within a dominance
hierarchy.

Is there such a role for spraints in otter societies, or are they merely
the result of elimination, the otter spending a long time in favoured
places, with the accumulation of spraints? Spraints may occur singly,
but often they are clustered into spraint piles, and a number of spraint
piles may be clustered into spraint stations. Some areas may be heavily
marked, such as the entrance to a holt or at rolling places, where otters
dry and groom themselves (Erlinge, 1967b). Otters frequently urinate
at sprainting sites. Sprainting sites are highly traditional and the
vegetation is frequently modified by regular usage. They are not
exclusive to individual otters. Sprainting places are described in more
detail in the next chapter (p. 47).

Erlinge (1968b) noticed that otters visit and examine signal spots
marked by other individuals, suggesting that information is being
imparted at the sprainting station. To test whether individuals recognize
and react to the spraints of strange otters, we carried out a small
experiment at Loch Broom in December 1980. Sixteen spraint piles were
located along the north shore and were marked with a drop of cellulose
paint. Eight piles were then chosen randomly and a spraint from a
captive adult male, from the Otter Trust, was placed adjacent to the pile.
After four days, only three of the eight control piles had been marked
with single spraints, while all of the eight experimental piles had been
marked with a total of 14 spraints. Otters, then, appear to respond to
the presence of foreign spraints by marking their spraint piles.

Trowbridge (1983) found that a captive male otter could discriminate
between the spraints of different individuals, even when both animals

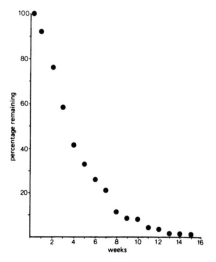

Fig. 2.12. Rate of disappearance of marked spraints along a Shropshire river in 1978. The number remaining with time can be described by the equation $S_t = 155 \cdot 3\, e^{-0 \cdot 34t}$ $(r^2 = 0 \cdot 96, \quad n = 92)$, where S_t is the percentage of spraints remaining at time t. (Authors' unpublished data.)

were previously unknown to him. Otters are capable of individual recognition using only spraint odour.

Trowbridge (1983) investigated the volatile constituents of the spraints of three otters, using the technique of gas–liquid chromatography. Each sample had a total of 98 peaks and the spraint profiles from individuals were relatively stable over the 25 days of study. Significant differences were present in the peak proportions between individuals, and otters would probably be able to recognize these differences. It remains to be seen how stable are the patterns in wild individuals, and indeed whether the technique can be used to determine the number of otters in the wild. The scents within the spraint are likely to derive from both the diet and the anal glands. The anal glands of the otter discharge via ducts to a bare glandular area just exterior to the anus, though they are much less specialized than in other mustelids (Stubbe, 1970). Otters also have proctodeal glands that open into the rectum. Gorman, Jenkins and Harper (1978) suggested that jelly-like deposits from the anal sacs of both male and female were related to the sexual cycle of the female, but Trowbridge (1983) showed that the origin of this 'anal jelly' was the gut and its production had no periodicity.

The effectiveness of spraints as signals will be enhanced if they persist for a long time. Jenkins and Burrows (1980) marked spraints and found that the rate of loss was linear with time, 50% of spraints disappearing in about two weeks. We have carried out a similar experiment on a Shropshire river, marking individual spraints with a spot of cellulose paint and recording their disappearance with time (Fig. 2.12). The pattern of disappearance, unlike that of Jenkins and Burrows, was exponential, i.e. a typical decay curve. Some 50% of spraints had disappeared in just over three weeks, with 90% gone in eight weeks. Nevertheless, some spraints, deposited in sheltered places, may persist for up to a year. It is clear that an otter may need to visit all parts of its home-range regularly if it wishes to maintain the impact of its signals.

There appears to be a sexual difference in the level of sprainting at signal sites. Green *et al.* (1984), collecting radio-active spraints from two females, considered that the level of marking was more than five times less than the young male followed by Jenkins (1980) or males labelled on the Outer Hebrides (Jane Twelves, unpublished data). In captivity, males were observed to spraint about seven times per active hour, whereas females sprainted only three times per active hour (Hillegaart, Ostman and Sandegren, 1981*a*).

There are also seasonal variations in the use of sprainting sites. Erlinge (1968*b*), in Sweden, found intense marking with spraints during the period October to March, when two year old males attempted to set up territories, followed by a fall in marking activity to a low in June and July. A similar pattern is found on the River Severn in Wales (Fig. 2.13), with a large winter increase in both sprainting sites and the number of spraints deposited. Watson (1978) found that many sprainting sites and trails that were well used in January were hardly used at all in April. However, this pattern may not be universal. Jenkins and Burrows (1980) recorded summer peaks of sprainting in two years on their study lochs in Scotland, followed by a year when there were no peaks. A very high peak of sprainting then occurred in the autumn. They attributed these peaks to the presence of family groups. Figure 2.14 shows the sprainting activity on the River Clun, Shropshire, over a six-year period. Several lines of evidence have indicated that the river was inhabited by a single female, who received periodic visits from a male living on the River Teme, below the confluence with the Clun. Marking tended to be heaviest in summer, but the pattern was not consistent. After the low level of marking in the winter of 1978/1979, the female was accompanied by a cub in the spring.

Not all parts of the otter's range appear to be marked equally. The Clun female deposited most spraints in the central area of her range. Green *et al.* (1984), collecting spraints labelled with ^{65}Zn, also found that females deposited their spraints in areas of high activity, rather than at the boundaries of the range. Captive female otters also sprainted more frequently close to their resting and grooming places (Hillegaart *et al.*, 1981*a*).

Male otters appear to mark the boundary zone of their ranges more

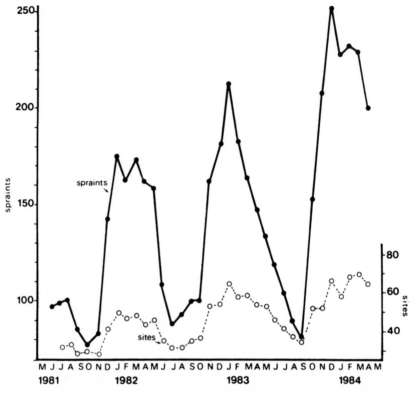

Fig. 2.13. Total number of spraints and spraint sites (expressed as three-point moving averages) recorded along one bank of nine study stretches (each of 1 km) along the River Severn, Wales, May 1981–May 1984 inclusive. (Authors' unpublished data.)

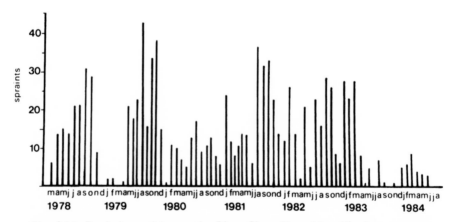

Fig. 2.14. Sprainting activity on the River Clun, Shropshire. The bars represent monthly summations of new spraints, recorded at weekly intervals, at 12 sprainting sites along 25 km of river between March 1978 and August 1984. (Authors' unpublished data.)

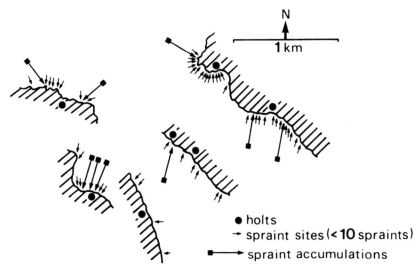

N

1 km

● holts
→ spraint sites (< 10 spraints)
●——→ spraint accumulations

Fig. 2.15. Position of spraint sites associated with holts at Loch Broom, north-west Scotland. (Adapted from Macdonald and Mason, 1980.)

intensely (Erlinge, 1968*b*). Trowbridge (1983) considers that Erlinge's data have been misinterpreted by later workers (e.g. Kruuk and Hewson, 1978; Macdonald and Mason, 1980) and there is not an increase in marking intensity at the boundaries. Nevertheless, Erlinge (1981) reiterates that he found intense marking along zones of overlap between males' ranges on inland waters. Green *et al.* (1984) also record a high level of activity in their radio-tracked males at the boundaries of their ranges, while Hillegaart *et al.* (1981*a*) found that captive males spent more time and sprainted more often in border areas.

The situation relating to marking of ranges appears somewhat different on the coast. Kruuk and Hewson (1978) found that more than 50% of spraints, in their Inverness-shire study area, were within 50 m of holts, that were regularly spaced with an inter-holt distance of 1·1 km. Jenkins and Burrows (1980), working in Wester Ross, found that holts were regularly spaced and about 0·75 km apart, while 73% of spraints were within 100 m of a holt. Macdonald and Mason (1980), working in winter at Loch Broom, found an average inter-holt distance of 1·1 km, though holts were situated in areas of suitable terrain and were not evenly spaced. Holt entrances were heavily marked with spraints, while there were large accumulations of spraints between holts, 62% of sprainting sites being within 150 m of a holt (Fig. 2.15). In South Uist, Outer Hebrides, Jane Twelves (unpublished data) found 73% of spraints within 40 m of a holt.

Trowbridge (1983) examined a 16 km length of coastline in western Scotland during the summer and found that spraint piles were clumped into stations, i.e. places of interest to the otter. Over 80% of all stations contained fresh water and spraint piles were also aggregated at rolling places and sleeping places. Spraints thus appeared to reflect centres of activity within home-ranges. Sprainting places were spaced regularly around the coast at a mean distance of 60 m. This was very different to the situation at Loch Broom, where long stretches of flat coastline, unsuitable for holts, were marked at a very low intensity. Trowbridge considered that the regular spacing of spraint sites was to ensure that an otter landing from the sea was never more than 30 m from a spraint site, thus maximizing its chances of encountering spraints.

Trowbridge's results may not be so much at odds with previous observations as they at first seem. Otters deposit spraints at regular intervals, with concentrations of spraints accumulating at foci of activity. For a male, one such focus of activity may be the boundary of a home-range. The home-range boundary is likely to be more definable in a linear riverine habitat, than on the coast, hence the observations of Erlinge (1968b) and Green et al. (1984). Other foci of activity for both sexes may be grooming places, washing places, playing areas and resting areas. Feeding areas may also be marked. In Portugal, heavy marking was found at the edge of pools containing fish, in rivers drying out in the summer heat (Macdonald and Mason, 1982b), and may have represented the defence of a dwindling resource.

Spraints, then, are deposited in prominent positions throughout the home-range of an otter, but with concentrations at centres of activity. Males appear to spraint more often at sprainting places than females. Spraints are undoubtedly used for communication, but can we say what information is being imparted? We can divide information into sexual and social. Otters frequently urinate on to their sprainting sites, so that information on sexual condition, such as the receptivity of the female, might be imparted by hormones within the urine. According to Trowbridge (1983), the main function of scent marking, using spraints, is to provide information by which individual distance between animals is maintained, so that conflict can be avoided. This is similar to the general hypothesis developed by Gosling (1982), who considered that marking by animals provides an olfactory association between the resident and its defended area, which allows intruders to identify the resident when they meet and thus reduce the frequency of escalated agonistic encounters. Intruders will compare the scent of any animals they meet with the memorized scent of males they have encountered in the vicinity. When the scents match, the resident is identified and the intruder withdraws. An animal that can defend an area long enough to mark it fully is likely to be of high physical quality and win most encounters, whereas intruders will vary in quality.

Trowbridge (1983) considers that otters are not territorial because their home-ranges are too large to be economically defendable. The males, at least, have an absolute dominance hierarchy and information concerning social position is conveyed by the spraint odours. This suggestion of a complete absence of territoriality seems inconsistent with the evidence available. At the same time, there is no evidence to suggest that rigid territoriality occurs. The answer probably lies somewhere in between and the social system is likely to be flexible in different areas, depending on particular environmental conditions and levels of resources. Radio-telemetry, should, in future, enable us to understand better the social system of the otter, but it will be necessary to radio-track several otters, of different social status and sex, concurrently, rather than consecutively, as has been done hitherto.

Activity

Over most of their range, otters are largely nocturnal, probably due mainly to disturbance and persecution. However, in some of the wilder areas, and especially on northern sea coasts, otters may frequently forage by day. Watson (1978), in coastal Shetland, observed that an adult female and juvenile showed variation in the periodicity of their activity, but there did not seem to be a consistent diurnal rhythm. Other otters on Shetland were consistently more nocturnal, which may have been due to human disturbance within their home-ranges. The tidal cycle may influence the activity of coastal otters, though Watson found them active at all states of the tide. In Norway, Lightfoot (1981) found that otters were largely nocturnal in autumn, but foraged during the day in winter, when they were dependent on the incoming tide breaking up the ice that formed along the sea's edge.

The radio-telemetry study of Green *et al.* (1984) provides information on activity and probably reflects the general situation in freshwater habitats. Otters were found to emerge at dusk, usually just after rather than just before sunset. Their retirement was less closely associated with dawn, taking place up to two hours before, but rarely more than 30 minutes after, sunrise. Some daytime activity occurred, but it never amounted to more than 6% of the total daily activity recorded. No activity was recorded in the afternoon.

Overall, otters showed two periods of high activity during the night, separated by a period of relative inactivity. The first activity period was the longest, beginning at sunset and lasting for three to five hours. The second period was more variable. There was a considerable amount of variation around this general pattern on individual nights. In general, the male otter studied was much more active than the female.

This radio-telemetry study has also provided information on the distance travelled by individual otters in a night. Otters use a number of resting places, with sites being changed daily on about 80% of occasions. The average distance travelled between resting places by two females was 1·0 km and 2·5 km respectively, but the total distance travelled was normally much greater than the distance between resting sites. Maximum nightly travels were as high as 3·8 km and 8·9 km for the two females. For the radio-tracked male the average distance between resting sites was 3·8 km, but his longest recorded nocturnal journey was 16·2 km. The females were observed to move rapidly between feeding areas, where they spent most of their time. By contrast, the male made longer journeys away from his activity centres, spending much time in the boundary zones of his range. Green et al. (1984) suggested that the male showed a cycle of increasing travel distances, there being intervals of approximately four days between the longest journeys.

These diurnal patterns of activity are likely to be superimposed on a seasonal pattern, but the data obtained by radio-telemetry have not been extensive enough to determine seasonal movements. In Sweden, Erlinge (1967b) found that the break-up of lake ice during the spring resulted in great activity of otters, abandoned feeding grounds being re-occupied and traditional sprainting spots being re-marked. From the middle of May onwards, the otters stayed in restricted areas for long periods, the males travelling less extensively. From late August, the activity of otters gradually increased, with home-ranges being fully exploited, but when the lakes froze in winter activity again became restricted to streams and other places with open water.

The climate is much milder in Britain and it is not known to what extent seasonal changes in activity occur, though seasonal variation in the level of sprainting (p. 33) suggests that there is a seasonal activity cycle. It is known that otters will move to headwaters and upland marshes in spring to prey upon spawning amphibians and they will also follow runs of salmon and sea-trout upstream.

Breeding

The female European otter appears to be continually polyoestrous, that is, there is a continuous oestrous cycle, with no specific breeding season. This seems to be the general situation in otters (Duplaix-Hall, 1975), only *Lutra canadensis* not conforming, by exhibiting delayed implantation (p. 166). According to Wayre (1979a), the bitch comes into oestrus every 40 to 45 days and remains so for about 14 days. Gorman *et al.* (1978) suggested an oestrous cycle of 30 to 40 days. Trowbridge (1983) collected urine samples daily from a captive female otter over a two-year period and measured the concentration of the hormone oestradiol. Peaks in the hormone concentration indicated a mean length of the oestrous cycle of 36 days, but with a range of 17 to 51 days over the period.

It is only for the purpose of mating that male and female otters have close contact; adult dog otters are otherwise largely solitary. Mating behaviour in captive otters has been described in detail by Harper and Jenkins (1981) and Pechlaner and Thaler (1983). Mating may take place on land or in water. Green *et al.* (1984) 'listened in' to the courtship activities of their radio-harnessed male. The otter pair showed vigorous activity, with chases through the undergrowth and intense vocalization. The male would also slap his tail on the ground. Both animals were so involved in courtship that they ignored passing vehicles and the close proximity of the observers. For about five days after the presumed mating, the male restricted his movements to the vicinity of the female, but without apparently making direct contact with her. Thereafter his travels returned gradually to normal.

Bitch otters are likely to choose the most secure holt within their range in which to give birth. Holts are lined with grass, reeds, twigs and other vegetation and the female adds substantial quantities of fresh material shortly before she gives birth.

The gestation period is about 62 days (Cocks, 1881; Corbet and Southern, 1977). Wayre (1979a) puts it somewhere between 61 and 74 days. On the birth of her young, the bitch ceases to mark traditional sprainting places; she becomes much more secretive and spends less time on land, presumably so that she leaves little scent by which she and her offspring could be detected by enemies (Green *et al.*, 1984). Hillegaart *et al.* (1981b) found that their captive females became aggressive to, and dominant over, males on the birth of their young. In the wild, males will not be in such enforced close proximity, but Green *et al.* (1984) suggested that their radio-harnessed male was aware of the birth of his cubs and tended to concentrate his activity in the vicinity of the breeding

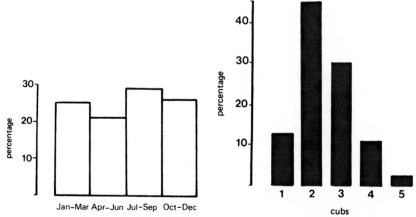

Fig. 2.16. Seasonal distribution of birth of cubs in Britain. Data from various sources ($n = 240$).

Fig. 2.17. Litter sizes of otters. Data from various sources ($n = 160$).

den, though it is not known whether direct contact was made with the female and there seemed to be no long-term parental commitment on the part of the male. It is generally considered that the male plays no part in the rearing of the cubs.

Being polyoestrous, the female otter can have cubs at any time of the year and, in Britain, this is generally considered to be the situation (Stephens, 1957; Harris, 1968). A distribution of cub births by season for Britain is shown in Fig. 2.16, which confirms this pattern. Nevertheless, several workers have suggested peaks in births. In the Western Highlands and in Shetland, most otters appear to be born in summer (Rowbottom and Rowbottom, 1980; Watson, 1978) though Jane Twelves (unpublished data) in the Outer Hebrides records year-round births. The summation of Fig. 2.16 may conceal regional variations.

On the Continent, Erlinge (1967b) reported that all of his otters in Sweden were born in spring, while Danilov and Tumanov (1975) reported breeding confined to spring and summer in the U.S.S.R. There was also a spring peak in the data collected by Reuther (1980a) for Niedersachsen, West Germany. In the marshes of Iraq, Thesiger (1964) stated that otters bred as early as January, but more often in February and March.

Although the otter can have young at any time of year, there may be a bias towards those seasons where the likelihood of successful rearing is high. In northern and continental climates, these seasons are likely to be the spring or summer, whereas year-round births may be the norm in western and southern areas. Jenkins (1980) has suggested that north-east Scotland may be a transitional area, with year-round births, but heavy cub mortality in hard winters. The data on birth dates are clearly inadequate for making generalizations: there is no information from southern Europe, for instance. Field workers should always note the appearance of cubs and measure their footprints, enabling an approximate back-calculation to the month of birth. The collation of this information over the next few years may help us to determine regional trends in breeding seasons, information of great value to the conservationist.

Litter size appears to be generally two to three young, with four to five young produced occasionally (Harris, 1968). Two appears to be the predominant litter size (Fig. 2.17). In East Germany, litter sizes averaged 2·3, in West Germany 2·8, in the Netherlands 2·8, and in Czechoslovakia 2·4, in which country a litter of six was reported (Stubbe, 1977; Reuther, 1980a; van Wijngaarden and van der Peppel, 1970; Baruš and Zedja, 1981). The average litter size produced by captive animals at the Otter Trust is 2·1 cubs. There are insufficient data from which to examine regional or seasonal trends in litter size, though such information would be extremely interesting.

At birth, cubs have short, pale grey fur and are about 12 cm long. They open their eyes when 30 to 35 days old, at which stage they can crawl. At one month old, cubs weigh 700 to 800 g, increasing to 1075 to 1250 g at two months old, at which time they are beginning to run and take solid food (Fig. 2.18). The cubs stay within their holt for about two to three months, when they emerge and take their first swim. It is often stated that the bitch has to teach her cubs to swim, but Wayre (1979a) considers there is much individual variation, both with cubs and bitches. Some cubs take to the water without hesitation.

At the age of four months, the cubs accompany their mother on hunting expeditions and become proficient at catching fish. This is also the time when otters are very playful, using extensive rolling areas on the land and playing with their prey. Sprainting activity also increases markedly.

The cubs stay with their mother for about a year and contact is very close throughout this period (Erlinge, 1967b). In Shetland, one cub was observed to stay close to the mother until 13 to 14 months old (Watson, 1978). The result of this long period of dependence means that a female can have no more than one litter of cubs per year and the interval between litters may very well be greater. Erlinge (1968b) suggests that, in Sweden, otters may breed only every second year.

There is virtually no information on the fate of cubs. Jenkins (1980)

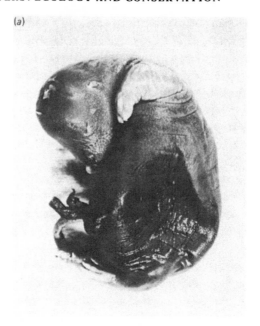

Fig. 2.18(*a*). For legend see p. 44.

considered that about 18 young otters were produced from 17 families over four years in his study area in north-east Scotland, suggesting considerable cub mortality. The average number of young produced in the river system was 0·5 young per 10 km per annum. Food shortage in hard winters appeared to be the chief cause of cub mortality. In East Germany, Stubbe (1969) determined the age structure of the otter population by examining skulls and teeth. He found that 42% of otters die within their first year, 33% survive for two years and only 25% live longer than two years. The period immediately following independence from the mother is clearly a critical period for young otters, as it is for most carnivores.

Female otters are physiologically capable of breeding in their third year, but some dog otters mature earlier. Wayre (1979*a*) reports a dog siring cubs at the age of only 18 months.

Vocalization

The European otter, being largely solitary, is a rather silent animal. Social otters are much more vocal. The significance of the sounds uttered by otters is not fully understood.

Fig. 2.18(*b*). For legend see p. 44.

Perhaps the most well-known call is the whistle, a piercing note of about one second's duration and carrying for some distance in still conditions. It is used by adults and by cubs from about two months of age and is a contact note.

The cubs begin calling at a very early age, uttering a bird-like,

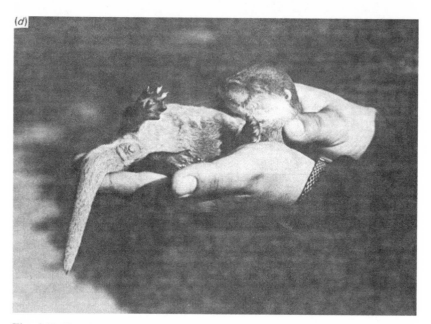

Fig. 2.18. Development of otter cubs: (a) foetus; (b) 1 hour old; (c) 3 days old; (d) 10 days old. (Photos (a) Linnet Barnes; (b), (c) Alpenzoo, Innsbruck; (d) Frischauf-Bild, Innsbruck.)

twittering call while searching for the mother's teats (Wayre, 1979*a*). When they emerge from the holt, they will squeak loudly from initial fear of the water. Watson (1978) described how a cub, temporarily deserted by its fishing mother, squeaked very loudly, the call being heard up to 400 m away. The calls became more frequent if the cub was left alone for longer, but after 10 to 15 minutes they became erratic.

Other calls uttered by otters include a loud 'hah', when the animal is startled, a 'whickering' greeting note and a threatening 'chittering' note, which may turn to a scream of rage.

Otters are also noisy in courtship. Green *et al.* (1984) observed sexual chases which were accompanied by staccato grunts and squeals. When the chases terminated, musical whistles and croons came from the undergrowth, which may have signalled copulation. On a second occasion, when the female appeared firmly to rebuff the advances of the male, loud screams, growls and chittering were heard.

Conclusion

On reading this chapter, it might appear that we know rather a lot about the ecology of the otter. However, with the possible exception of diet, the information we have is very sketchy; it is based on relatively short-term studies of very few animals from restricted geographical regions. Much of the research is very recent and has been largely stimulated by a decline in populations of otters over wide areas. The current status of the otter and ways of investigating it are the subject of the next chapter.

3 DISTRIBUTION AND STATUS

Thyself, and melancholy streams

RICHARD LOVELACE,
To the Grasshopper

The otter's nocturnal and elusive habits make direct counts of animals almost impossible. In assessing distribution and status, biologists have frequently resorted to information from hunters and fishermen, but there are problems in interpreting such data (p. 55). Standardized field surveys based on searches for signs of otters should provide more reliable results.

Otter signs

The sign most commonly found is the otter's dropping or spraint. When fresh, spraints are black and have a pleasant, sweet-musky odour which fades with time. The colour is also gradually lost, the spraints drying to become grey and eventually, white. The size and shape vary from a tiny blob or tar-like dribble to a compact cylindrical dropping ten or more centimetres long (Fig. 3.1). When otters have been eating snakes, the spraints may appear very large because long lengths of snake skin are passed. (Those containing snake can also smell most unpleasant!) More usually, spraints contain fish bones and scales, amphibian bones or fragments of crustacean exoskeleton. Fur and feathers may occasionally be present.

Spraints are often deposited at conspicuous sites (Figs. 3.2, 3.3). They may be found on large boulders and logs in the water or on the banks, in hollows under eroded tree roots, on concrete or stone under bridges or at the confluence of streams. Where otters use an overland route, perhaps to cut across a river meander, the path may be marked with

Fig. 3.1. Otter spraints: (a) a fresh spraint on a stone; (b) spraints deposited on pebbles under a bridge. The arrow indicates a tar-like spraint frequently deposited by otters.

spraints at either end and also along its length. On the coast, many paths lead to small freshwater pools where large quantities of spraint accumulate (Fig. 3.4).

Many sprainting sites are traditional, being re-marked frequently by the same animals and by generation after generation. On grassy banks the results of regular marking are clearly visible as the lush tufts of greener nitrophilous grasses can be spotted from several metres away. Figure 3.5 shows two such sites – on the Scottish coast and on a Greek river.

In Wales, on elevated treeless banks, we have found larger areas which are intensively marked during the winter. These areas of grass, up to 2 m × 1 m in size, are trampled, scratched and smeared with spraints and anal secretions. The vegetation quickly turns brown, making the site conspicuous between about November and March. During the summer, marking is irregular and the grass gradually recovers only to be scraped up again the following winter. Erlinge (1967b) described similar areas as 'rolling places', which he also found most frequently in the autumn and winter, but in Sweden they were often hidden under willows. We have not found this type of marking behaviour on rivers holding very low otter populations.

Otters also scrape up mounds of sand or vegetation, frequently depositing spraint or urine on top (Fig. 3.6). Usually such mounds are small, but one heap of grass found under a bridge was enlarged and sprainted on all summer until it resembled a football. Jim and Rosemary

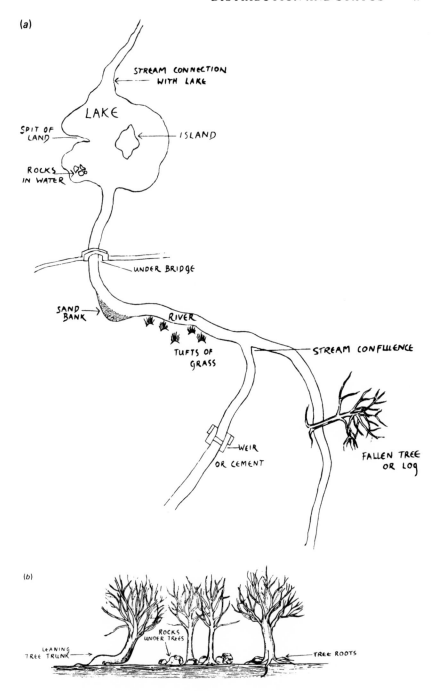

Fig. 3.2. Diagrams of a lake and river showing typical sites where otters deposit spraints.

Fig. 3.3. Typical sprainting sites: (a) a prominent boulder at the edge of a river; (b) a log washed up on the bank; (c) the entrance to a cavity under the eroded roots of a sycamore tree; (d) ledges under a bridge where spraints may be deposited at two levels, as indicated by the arrows. The higher ledge is at a height of 1·3 m above the lower ledge.

Green (pers. comm.) discovered a similar object in Scotland. Why otters scratch up earth or vegetation is not known, but such signs seem to be a feature on river reaches holding high populations. Erlinge (1967b) associated this behaviour with the otter's preference for sprainting on elevations, but also recorded much scratching at the boundaries of home-ranges. He suggested that dog otters scratched the ground more often than females or cubs. On the banks of Greek rivers, sand mounds made by otters are common and we observed groups of up to six along 3 m stretches of bank. Other animals, like the red fox, also tend to deposit

Fig. 3.4. This freshwater pool on a rocky coast in western Scotland is regularly used by otters. The small mounds at the far edge of the pool have been formed by the deposition of spraints over many years.

droppings on elevations and Hornocker (1969) records mountain lions (*Felis concolor*) scraping up vegetation before urinating on it.

Otters also spraint at the entrances to holts (underground dens) or at other resting places although, as Green *et al.* (1984) discovered, by no means all sleeping sites are marked. Sometimes, however, a clear path leads from the water to cavities in the eroded roots of a bankside tree, where spraints and even remnants of bedding may be found. On the Scottish coasts, holts among rock are usually heavily marked with large piles of spraint at the entrances (Kruuk and Hewson, 1978; Macdonald and Mason, 1980).

Any resting place located above ground was described as a couch by Green *et al.* (1984), whereas to Hewson (1969) a couch was like a platform constructed of vegetation that became hollowed with use. Green and Green's (1980) account of their survey in Scotland illustrates a couch on an upland marsh which resembles a bird's nest, and Harris (1968) records observations of otter constructions like coot nests. Hewson noted spraints close to otter couches and we found them on a reed platform by an East Anglian river. In Greece recently, we saw a clear otter path leading up a steep bank to a scraped-out hollow under alders. The bowl-like cavity contained no bedding but there were

Fig. 3.5. Regular deposition of spraints can result (*a*) in a prominent mound (in western Scotland) that gradually becomes grassed over. At traditional sprainting sites, tufts of nitrophilous grasses may flourish. On this stretch of the River Kuru in Greece (*b*), each of numerous small islands bore these tufts, which were avoided by grazing sheep. Such signs of otters can be spotted from a distance of many metres.

Fig. 3.6. Small piles of (*a*) dead grass or (*b*) sand are scratched together by otters and frequently sprainted upon. The significance of this type of behaviour is not understood.

spraints around the rim. So, even without the aid of radio-telemetry, at least some resting sites can be found by simple field observation.

In addition to spraints, otters also produce jelly-like secretions and the source of these is discussed on p. 32. Some secretions may be just gut mucus. Pat Foster-Turley (pers. comm.) found that her captive *Aonyx cinerea* sometimes excreted large quantities of green mucus and we have observed this in the wild in Britain. Secretions may also be brown, black, yellow or white and, during monthly monitoring on the River Severn in Wales, it became clear that the deposition of secretions increased in the winters.

Published guides to the tracks and signs of mammals, e.g. that of Lawrence and Brown (1967), often mention the otter slide, which is usually thought of, perhaps rather sentimentally, in relation to cubs at play. In snow, the trails of otters sliding down slopes or over ice are clear, but such behaviour may be less about play and more to do with conservation of energy. Erlinge (1967*a*) never found 'tobogganing spots' in Sweden and slides may often be simply those places where the animal habitually enters or leaves the water.

Mud and sand reveal the footprints of the otter. These are easy to recognize when all five toes are visible (see Fig. 3.7). In soft substrata, the claws may show but the webbing is seldom seen. Confusion with tracks of other mammals may occur when, as so often happens, only four toes show clearly. In Britain, confusion is most likely to occur between

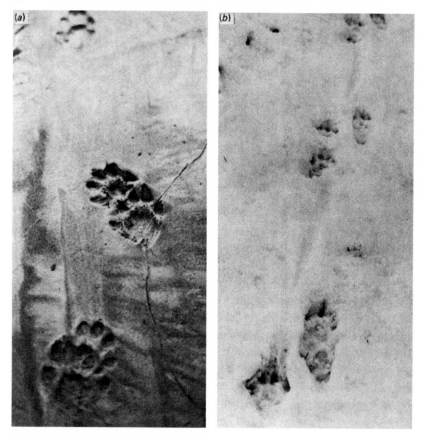

Fig. 3.7. Otter tracks (*a*) in mud under a bridge; (*b*) in snow. The imprint of all five toes may not always be visible. Note that the track has a rounded appearance.

signs of mink and of otter, and Fig. 3.8 illustrates the tracks and scats of mink for comparison.

Occasionally, it may be possible to track an individual otter from its prints. Amy Lightfoot (pers. comm.) could follow an animal on a Norwegian fjord because damage to its toes gave it distinctive tracks and much of Erlinge's work in Sweden was based on his recognition of individual prints. The size of the track can also give a very rough guide to the kind of animal present. If measured across the widest part, to include all five toes, it may be estimated that a track of over 7 cm probably belongs to a male and one of less than 4·5 cm may be a cub. However, the size of prints can vary widely with the softness of the substrate and there is no way of distinguishing prints of females from

Fig. 3.8. Field signs of mink (*Mustela vison*). (*a*) The footprint is more stellate than that of the otter; (*b*) a mink scat on a pebble. The dropping of the mink is usually more compact than that of the otter and often appears cork-screwed. Often it will contain fur or feathers and, if fresh, smells vile!

those of juvenile males. Tracks have been found quite commonly in Greece that measure 9 cm and it may be that geographical variations limit the usefulness of print measurements.

Food remains are commonly cited as sure signs of an otter's presence, but any half-eaten fish by a river should not be assumed to be the remains of an otter's meal unless spraints or prints are found associated with it. Partly eaten fish may be the work of foxes (authors' pers. obs.), crows or other scavengers.

Given that otters are so rarely seen, the most certain evidence of their presence is spraints and clear prints. These are the signs that are searched for in field surveys to-day.

Methods of survey

Traditionally, information on otter distribution was collected, not in the field, but through questionnaires circulated to hunters, fishermen, game wardens or naturalists. Based on such methods, distribution maps have been produced in, for example, Holland, East Germany and Scandinavian countries. The reliability of questionnaire returns is, however, open to much doubt. The method is dependent upon the accuracy of memory and results may reflect the distribution of interested contributors rather than that of the animal. Those with a vested interest in the otter, such as hunters or fishermen, may provide information suited to their own requirements and, in areas where the otter is known to be a protected species, locals, either because they care or snare, may be reticent about giving information.

Occasionally, valuable data may be gleaned from questionnaire returns but they should always be treated with extreme caution. Cassola (1980) cast doubt on the results of such a survey published by Cagnolaro *et al.* (1975) in Italy and, now that field surveys have been carried out, it is clear that his doubts were justified. Similarly, in Spain, Delibes and Callejo (1985) found that Elliot's (1983) field survey provided a truer picture of otter distribution than the more optimistic results from questionnaires reported by Blas-Aritio (1978).

Another way of collecting information on distribution and status involves the analyses of hunting returns. In the United States, skinned carcasses of *L. canadensis* supplied by trappers are used to give information on the age structure and reproductive status of local populations. With thousands of dead otters to examine, the U.S. Fish and Wildlife Service can collect data on which to base adjustments to annual harvest quotas.

In Britain, sport hunting records have been analysed. At one time there were 13 packs of otter hounds operating in Britain and the huntsmen kept records of the number of otters found. By comparing the success rates of the different hunts over several years, patterns relating to otter abundance could be obtained. In 1962, Lloyd suggested that several of the British hunts were finding fewer otters and in 1969 the Mammal Society published a report based on hunt returns (Anonymous, 1969). A decline in otter numbers appeared to have occurred in many parts of England and Wales and a comparable report in 1974 (Anonymous, 1974), suggested further declines in the English Midlands with little sign of recovery elsewhere. Results at this time from the single Scottish hunt indicated no reduction. These reports certainly pointed out that the otter was becoming rare in much of Britain but they did not tackle the real problems of actual distribution or abundance. They also ignored the biases that can arise from such data.

Chanin and Jefferies (1978), who used the British hunt returns to determine a more precise timing for the initial decline, discussed some of the problems in interpretation. While individual packs mostly hunted their own areas, they also exchanged visits with other hunts and all the results were recorded together. Thus a pack with poor returns for its home area could inflate its success rate with records of productive visits elsewhere. As otters became rare, some hunts called the hounds off before the kill but still recorded drags, i.e. the otter was scented but not found. However, if otters are more mobile when at low population densities, then records of drags may not be comparable to records of finds. It is also clear that, as the animal became rare, hunts tended to concentrate on those rivers offering better chances of success and such selectivity will also lead to optimistic biases in the records.

The number of days spent hunting also varied from year to year. Since the records represent the number of finds per hundred days' hunting, even a small variation in the number of finds can lead to misinterpretation of results in years when the number of hunt days was low.

Hunt data must, therefore, be analysed with care and cannot be accepted as providing definitive results. They can, however, as demonstrated by Chanin and Jefferies, point to population trends. These authors used the information to show that sharp declines in otter numbers occurred in the late 1950s and early 1960s. They then postulated that the most likely cause of this was the organochlorine pesticide dieldrin (see p. 92).

Before the 1970s, very little field work had been undertaken in attempts to estimate otter populations, although Stephens (1967) had combined information from hunts and from local people with some of her own field observations. She concluded that otters were common throughout most of Britain, including the Greater London area, and recorded them as scarce only on the Bristol Avon and the Mersey. Five years later, Lloyd, using hunt data, suggested that the situation may have changed.

In 1973, the Mammal Society initiated a field survey to be carried out throughout Britain by volunteers. The results suggested that the species was rare but not all areas were surveyed. The first systematic attempts to determine otter distribution were carried out in East Anglia (West, 1975; Macdonald and Mason, 1976). West reported results collected by many field surveyors, together with information from water bailiffs and staff of the Ministry of Agriculture, Fisheries and Food. He estimated 36 otters in Suffolk and suggested that numbers had declined steadily since the 1940s. In Norfolk, we searched the environs of all bridges over waterways named on the 1″ O.S. maps. Of 233 sites visited, signs of otters were found at 32 (14%), an additional 8 records being received from other naturalists. Based on otter range sizes given by Erlinge in Sweden, we estimated 17 otter ranges in Norfolk which seemed to be well below the carrying-capacity of a county rich in rivers, lakes and marshes. In retrospect, it was probably unwise to estimate the number of ranges since it is now clear that range sizes can vary considerably. However, the distribution of field signs did show clearly that the animal had become very rare.

By 1976, conservationists were finally beginning to show real concern about the status of the otter in Britain and the Joint Otter Group was formed to include members of the Nature Conservancy Council, the Society for the Promotion of Nature Conservation (now the Royal Society for Nature Conservation), the Mammal Society and the Institute of Terrestrial Ecology. The group produced two reports (Joint Otter Group, 1977, 1979) and, in the first of these, recommended that national field surveys be conducted to determine the real distribution of the otter in Britain. Surveys were carried out between 1977 and 1981.

Distribution in the British Isles

Full-time surveyors were appointed to carry out the national surveys and the work was completed in as short a time as possible so that the results represented the distribution of otters at a given time. The surveys were based on the 10 km square national grid, with sites visited throughout Wales and Scotland. In England and Ireland, alternate 50 km squares only were visited because of a shortage of time. In England the north-west and south-east quarters of each 100 km square were surveyed, while, in Ireland, the north-east and south-west quarters were visited. In retrospect it seems that in Ireland, because otters were found almost everywhere, the survey of alternate squares gave sufficient information. In England, however, where distribution proved patchy, some useful local information was lost.

To assess any future change in the distribution or status of the animal, it was decided that the surveys should be repeated at seven-year intervals and so standard methods were developed. Sites were visited, where possible, at 5 to 8 km intervals along rivers, lakes or coasts and were usually chosen for ease of access, e.g. at a bridge or where a road ran close to water (Macdonald, 1983). At each site, a maximum of 600 m was searched for spraints or footprints. As soon as signs were found the search was terminated and, if nothing had been found after 600 m, the site was considered negative. Spot checks at bridges other than chosen survey sites gave some additional information.

A data form was completed at each site, giving information on habitat and on some factors likely to affect otter distribution. Much of this information may, however, be of limited value. Certain features of a river such as width, depth and macrophyte growth vary with seasonal conditions and a description of a site made following a fleeting visit when spraints were found immediately may differ vastly from a description of the same site following a 600 m search. Thus, more environmental information can be gleaned from data forms recording negative sites.

During the surveys, most sites were visited once only and several factors can influence the reliability of data collected in this way. A few days of heavy rain or spates can wash away many spraints, while potential sprainting sites, like ledges under bridges, become submerged. Otters may move away in response to both floods and droughts, or leave an area to follow some rich, if temporary, food supply such as spawning trout or frogs. Sometimes they seem to desert an area for no obvious reason at all or may visit a stream without leaving any spraints (authors' pers. obs.). In addition, the animals may leave more or fewer signs according to seasonal variations in marking activity or to the type of habitat. The ease of finding spraints also varies with habitat. A featureless bank smothered in summer herbaceous vegetation can be very hard to search!

In areas with low populations, relatively few spraints may be found. During the national surveys stretches longer than 600 m were searched in order to assess the accuracy of the method. It was concluded that, in areas where otters were very scarce, the level of marking was often so low that their presence was overlooked. Claus Reuther (pers. comm.) also found that a 600 m search was insufficient in parts of West Germany, where the otter is close to extinction. Thus, the method will not detect every otter range but it should indicate clearly areas where otters are thriving and regions where populations are fragmenting.

Throughout the British Isles, 10 979 sites were surveyed and at 5942 (54%) signs of otters were found. For a country with a high human population and with advanced agriculture and industry, the overall result may seem surprisingly high, but there were large regional differences.

In Wales, the survey was carried out in 1977 and 1978 (Crawford, Jones and McNulty, 1979). Signs of otters were found at 210 (20%) of the 1030 sites surveyed and an additional 8% of 567 spot checks proved positive (Fig. 3.9). The species seemed to be absent from much of industrial south Wales and from areas in the north. There are 15 major river catchments in Wales and the highest percentages of positive sites were found on the Cleddau, Wye, Dee, Severn, Teifi and Ystwyth. The animal still appeared to be thriving in parts of mid-Wales, with a stronghold in the south-west of the country.

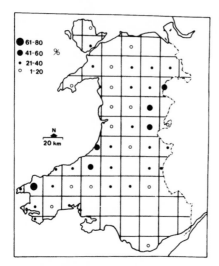

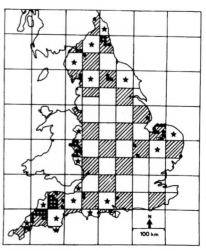

Fig. 3.9. Distribution of otters in Wales (1977–1978) as a percentage of positive results in 20 km square tetrads. No otters were found in the empty tetrads. (Adapted from Crawford *et al.*, 1979.)

Fig. 3.10. Otter distribution in England. The shaded 50 km squares were visited by Elizabeth Lenton during the national otter survey of 1977–1979 (Lenton *et al.*, 1980). Closed circles indicate positive 10 km squares, the open circles are positive spot checks. The stars show 50 km squares, not examined during the national survey, where otters have been recorded since 1977.

In England, surveyed between 1977 and 1979 (Lenton, Chanin and Jefferies, 1980), 2940 sites were visited, with signs of otter found at 170 (6%). The species was absent or very rare in much of central England – a sharp contrast to the work of Stephens, who, in 1957, had reported otters throughout the region (Fig. 3.10). In Elizabeth Lenton's survey, the highest number of positive sites was recorded in south-west England, with animals still surviving in Norfolk in the east, on the borders with mid-Wales and in the north (Northumbria, Yorkshire and Cumbria). Other scattered records indicated the presence of remnant populations or even individual otters, isolated and with poor chances of survival.

The results of the Scottish survey (Fig. 3.11) proved much more optimistic. Green and Green (1980) visited 4636 sites between 1977 and 1979 and found otters at 3385 (73%). In their report, however, they warned against complacency for, while the Highlands, islands and some coastal regions produced excellent results, it was clear that a decline had occurred in the eastern and central lowlands and in the southern uplands.

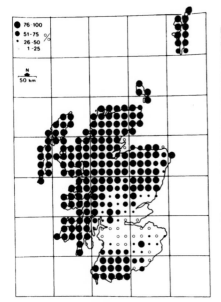

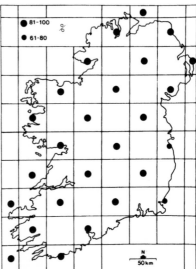

Fig. 3.11. Distribution of otters in Scotland (1977–1979), as a percentage of
positive results in 20 km square tetrads.
No otters were found in the empty tetrads.
(Adapted from Green and Green, 1980.)

Fig. 3.12. Distribution of otters in Ireland
(1980–1981) as a percentage of positive
results in 50 km squares. Blank squares
were not surveyed. (Adapted from Chapman and Chapman, 1982.)

The survey of Ireland by Chapman and Chapman (1982) produced
the highest results for the British Isles (Fig. 3.12). In 1980 and 1981 they
visited 2373 sites and recorded otter signs at 2177 (92%) of those sites.
It seems that, in Ireland, otters occur almost everywhere, although
analyses of the data did show that more positive results were obtained
from the western and central regions than from the east of the country.
Areas which produced fewer than average positive sites included the
Dublin area, the lower Blackwater/Bride system, the lower River Barrow
and the Burren.

The national surveys were of great value in giving a broad overview
of the situation of the otter in the British Isles and, at present, no similar
data have been collected anywhere else in Europe. We are also fortunate
in Britain in having many field biologists currently monitoring local otter
populations and their studies provide continuing information on changes
in status or distribution. In East Anglia, for example, the species appears
still to be in decline. West (1975) estimated 36 animals in Suffolk in the
early 1970s, while Lenton et al. (1980) found signs at one site only. Today
the natural otter population of Suffolk is very low, although captive-bred

animals have now been released there (see p. 154). In Norfolk, the field survey by Macdonald and Mason (1976) was repeated by Clayton and Jackson (1980), who reported a 25% decrease with signs no longer found on two major rivers, the Bure and the Yare.

The Norfolk population is now very isolated with little obvious chance of recruitment from elsewhere. The River Teme catchment, on the borders of England and mid-Wales, is less isolated but a continuing decline also appears to be occurring there (authors' pers. obs.). In 1977 (Maconald, Mason and Coghill, 1978), otters could be found on the upper reaches of the Teme and on three major tributaries, the Rea, Onny and Clun. The Rea, the furthest downstream of the tributaries, has not held resident otters since 1980, and regular marking has not been recorded on the Onny since 1982. The Clun, the most upstream of the tributaries, contained at least one resident otter until late in 1983, but subsequent marking activity was irregular, suggestive of occasional visits by animals from the main Teme. The otter was not common on the Teme catchment in 1977, but today the remaining numbers must be very small. It appears that they are now resident only on a short stretch of main river and make periodic excursions up the tributaries. Thus, judging by marking patterns, it seems that between 1977 and 1984, three-quarters of the area previously occupied by resident otters was largely deserted.

In south-west England too, there are areas that continue to lose their otters. The declines on the Somerset Levels are described later (p. 115) and Elizabeth Lenton (pers. comm.) has also noted recent and marked reductions of otters on Dorset rivers. In the survey of Exmoor National Park, Jarman (1981) concluded that, while the species had suffered losses in the late 1950s and early 1960s, further declines had also occurred since 1975.

Regional studies give valuable information on local populations but repetition of the national surveys should illustrate any general trends in distribution. Since it was decided that the British national surveys should be repeated at intervals of seven years, the second survey of Wales began in 1984. The initial results suggested an increase in distribution, with most river systems showing a rise in the number of positive sites. For example, the results for the Dovey area were 8% positive in 1977 and 34% in 1984 (authors' pers. obs.). Large increases in success rates were also recorded in the catchments of the Tywi, Ystwyth, Clwyd, Wye and Severn. However, caution is required in the interpretation of such results since one survey can never quite replicate another. There will always be variation in the competence of surveyors and variation in weather conditions. The River Conwy, for example, was surveyed after spates in 1977, when only one site produced signs of otters. In 1984, in conditions of very low water, 19% of the replicated sites were positive. Despite this increased success, the distribution of otters in the Conwy catchment is still limited and it may well be that no real change had occurred in the seven year interval.

On the Severn catchment, 39% of sites were positive for otters in 1977, 67% in 1984. The second survey was carried out after several months of dry weather which meant that very old spraints were still detectable. On the upper Teme (a Severn tributary), signs were recorded on the headwaters of the river. However, because the Teme has been monitored regularly, it is known that those same spraints had been deposited by an animal that visited the area three months previously. A few positive records of this type can dramatically increase the overall success rate for a catchment but may not reflect the normal distribution of resident otters.

Where populations are very low, animals may wander over large distances and the majority of signs located may be old, single spraints. With more viable populations, recently deposited spraints can be found more frequently and there is a greater likelihood of sites being marked with more than one spraint. Green and Green (1980) found the highest number of signs per successful site on the coasts of the Scottish Western Isles. They used the numbers of signs found at sites as an indicator of the quality of populations, relating their findings both to region and altitude.

Distribution in northern Europe

More detailed information is available on otter distribution in Britain than for any other part of its range. Many of the data from Europe have been based on game returns but now, with growing concern for the species, groups of biologists are carrying out survey work in many countries.

In Sweden, Erlinge (1972a) compared the numbers of otters killed during hunting seasons from 1938/1939 to 1968/1969. He showed that there had been a steady decline since 1950 so that, by 1958, the number reported shot was about half of that in the 1940s. During 1965–1966, in co-operation with the Swedish Hunting Association, information was gathered by questionnaires, and Erlinge (1972a) concluded that animals were generally scarce. Large decreases appeared to have occurred in some central and southern parts of the country and the species seemed to be uncommon in the north. Erlinge did, however, suggest that populations were higher on the Baltic coast south of Stockholm and on the west coast near the Norwegian border. Ten years later the survey was repeated but, despite protection, the otter had made no recovery within that period (Erlinge, 1980). The distribution map presented in 1980 does, however, show otters as 'locally common in some water systems' in the south-east and north.

Field surveys began in 1983 and initial results from central and southern Sweden suggested further declines (Olsson and Sandegren, 1985). The abundant lakes and *Phragmites* beds of these regions provide ideal habitat for otters, but the Swedes consider the main problems to be contamination by polychlorinated biphenyls (see p. 97) and general acidification, which restricts otters to eutrophic waters.

In Norway also, information on otter distribution has been collected from questionnaires and game reports. Myrberget and Fröiland (1972) illustrated a decline in numbers between 1965 and 1970, while Heggberget and Myrberget (1980) reported that little change occurred between 1970 and 1977. They considered that good populations still thrive in the coastal districts of Trondelag, with numbers increasing in the north. It seems that the south of the country holds fewest animals.

Magner Norderhaug, reporting to the 1983 International Otter Symposium in Strasbourg, suggested that the Norwegian populations may be amongst the best in Europe and he emphasized the importance of north Norway. Without concrete evidence he estimated 10000 animals for the country.

Within Finland, Pulliainen (1985) estimated 1000 otters, with viable populations in central, eastern and northern regions.

All estimates of actual numbers of otters must be treated with considerable reserve, but it seems clear that northern Scandinavia, despite substantial losses, currently holds more otters than do countries further south. In West Germany, Holland, Belgium and Denmark, for example, few animals remain.

In West Germany, otters have been in decline for the last hundred years. In his study of the species in Lower Saxony, Reuther (1980*a*) showed that numbers fell at the end of the nineteenth century and then declined again rapidly in the 1950s and 1960s. From his census in 1977–1979 he calculated 200 to 400 animals remaining and today only isolated and fragmented populations are left. In Bavaria, according to Claus Reuther (pers. comm.) there are probably fewer than 10 individuals.

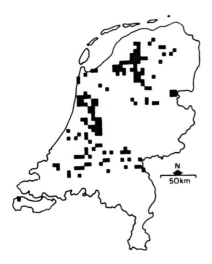

Fig. 3.13. Distribution of otters in the Netherlands. Black squares indicate 10 km squares known to contain otters during the survey period 1979–1983. (Adapted from Veen, 1984.)

Hodl-Rohn (1980) thought it probable that otters found in the Bavarian Forest were transients from Czechoslovakia and those animals now face new problems as water acidification limits their feeding sites to private fish ponds. Very recently, signs of otters have, however, been found in eastern Hessen.

Gunter Heidemann has been working on the otters of Schleswig-Holstein in north Germany for several years and has collected data since 1955. Between 1967 and 1969 animals were reported from 93 localities but from only 34 in 1977–1979 (Heidemann, 1980). He calculated a 46% decrease between 1967–1969 and 1970–1976 with a 50% reduction between 1975 and 1977–1980 (Heidemann, 1981). Now he can find evidence of the species at only 19 places and it is thought that less than 50 individuals still live in the region.

In Holland too, the distribution of otters seems to be patchy. Van Wijngaarden and van de Peppel reported in 1970 that numbers had increased from the estimated 30 to 50 of the early 1940s, but thought that viable populations were restricted to five isolated regions. In 1980, van Wijngaarden suggested that stable populations occurred in parts of Friesland, Groningen and Noord-Holland, with a few animals still in Utrecht and Zeeland. Veen (1984), from a census of otters carried out between 1979 and 1983, showed that they were more widely distributed than was previously thought (Fig. 3.13).

In 1970, van Wijngaarden and van de Peppel reported otters on the border lands with Belgium, but by 1980, according to Kesteloot (1980), only a few nomadic specimens remained in Belgium. Libois *et al.* (1982) illustrated a few scattered sites where animals had been recorded since

1975. They considered that, in Wallony, no more than 20 remained, while the situation in Flanders appeared equally bleak (Criel, 1984). On a short visit to northern Belgium in 1983, we could find no signs of otters. Many of the rivers were grossly polluted and only isolated remnants of suitable habitat seemed to remain. Descy and Empain (1984) consider that Belgium is one of the most backward countries in Europe, with regard to waste treatment. They report, for example, that only 6% of sewage receives treatment, compared with 80% in neighbouring Luxembourg. A field survey has now been initiated by the carnivore protection group, Nationale Campagne Bescherming Roofdieren, and a more accurate picture of otter distribution in Flanders may soon be available.

In Denmark, the situation of the otter is also precarious. Sharp declines have occurred since 1960 and in 1980 Jensen (1980) thought that the species could only be found on some of the islands and in part of Jutland.

Thus, throughout the central belt of northern Europe declines have been dramatic and, as usual, pollution and habitat destruction are cited as contributory factors. Unfortunately, however, little work has been carried out on levels of pollutants in otter tissues or in prey items, there have been few real assessments of habitat quality and only a small proportion of population estimates are based on field observations.

In France, recent field surveys have provided useful information on otter distribution. In 1980, Jim and Rosemary Green investigated 315 sites and found otter signs at 46 (15%). They located animals in the uplands of Brittany, the Massif Central and the Pyrenees and in the coastal plains of Landes-Gironde (Green and Green, 1981). They were puzzled by the apparent lack of more widespread populations, since much suitable habitat remained and they postulated that water pollution may have caused problems. In 1981, Christian Bouchardy produced an informative booklet on the otter and its field signs and he, together with other members of the French Groupe Loutres, have now surveyed many regions.

The otter was common in France until 1930, but today it is absent from large areas with high human populations or intensive agriculture. It now occurs in the west of the country, but to the east of a transect from Rouen to Montpellier it is very rare (Groupe Loutres, 1983). The species has been lost from the Paris region, and the few still surviving in Alsace in 1975 (Baumgart, 1977) had disappeared by 1978–1979. Some otters still live in the Alps, but there are not many in north-east or east-central France.

Lutra lutra is, however, widespread in the western Massif Central and Charentes, and from Brittany through Vendee and Aquitaine to the Spanish border. In the Pyrenees, they have been recorded at 2200 m! Braun (1984) suggests that the only true coastal otters in France are those found on islands off the coast of Brittany. Kempf (1985) found that, within those areas still occupied by otters, they were present at 30% of sites with suitable habitat but at only 0·3% of sites where habitat was considered poor. In the Camargue, which offers some of the best otter habitat in France, the animal has not been recorded since the 1950s (Luc Hoffmann, pers. comm.).

In Switzerland, the otter was, at one time, widespread and could be found on most waterways. At high altitudes, sub-optimal habitat for otters, the density was low but there they could avoid growing human pressures in the plains. However, by 1953 the Swiss population was considered to consist of around 150 specimens and by 1975 only about 12 were thought to remain (Müller *et al.*, 1976; R. Lebeau, *in litt.*). A reintroduction programme was instigated with eight otters from Bulgaria being released in Canton Berne (see p. 154).

In Austria, the species fares little better, being restricted to forest habitat in the Bohemian Massif and some foothills of the Alps (E. Kraus, 1980). In Luxembourg, the otter is probably extinct (Reuther, 1980*b*).

Many northern European countries, therefore, seem to hold little hope for stable otter populations but, in the south, some places still provide suitable conditions and some can boast good numbers of animals.

Distribution in southern Europe

Until recently, little information was available from southern Europe but, with otter numbers declining in the north, it became increasingly important to gather data. In August 1980, we visited central Portugal to test whether, in a very short period, enough field data could be collected to give useful guides to the status of the species. This type of survey has now been used several times in Europe and North Africa.

Only a few weeks were spent in any country, with as wide an area as possible being visited. No attempt was made to cover waterways evenly at 5 to 8 km intervals as in the British national surveys but, as in Britain, sites were chosen for ease of access. At each point, a maximum of 600 m was searched but we preferred to explore a minimum of 200 m even if spraints were found immediately, as this gives some indication of marking intensity in different habitats. A simple data form was completed at each site. Obviously this method of survey cannot provide as complete a coverage as was achieved in Britain but it makes it possible to identify regions where viable otter populations still occur and areas where they may be scarce. The method is also of value if results are needed urgently or financial support is limiting.

In central Portugal, we visited 90 sites and found signs of otters at 70% (Macdonald and Mason, 1982b). Animals were located on all types of waterway except at reservoirs. The highest level of marking (mean of 6·4 sprainting sites and 12·0 spraints per 200 m) was found on rivers that were so dry that only isolated pools remained. On rivers where summer desiccation was less severe, so that pools were interconnected by trickles of flowing water, the level of marking was significantly lower. Isolated pools held quantities of 'captive' fishes, providing easy access to food but, as deoxygenation proceeded, potential prey was lost. The high levels of marking may simply have indicated that otters were forced to congregate at pools or it may have been linked to increasing defence of a dwindling and vital resource.

Portugal offers exciting scope for research into otter biology. Over much of the country the animal's habitat must alter radically between summer and winter and the real effects of apparently high human disturbance on rivers and the presence of large numbers of feral dogs are unknown. Unfortunately, however, very few people are currently working on Portuguese otters. Ferrand de Almeida (1980) indicated a few rivers in central Portugal where otters still occurred and in 1983 Simões Graça and Ferrand de Almeida confirmed the presence of animals on the lower Mondego catchment, with the help of questionnaires sent to hunters and game wardens. In their study of the Marsh of Arzila, in the Mondego catchment, they expressed fears that illegal hunting for pelts could be reducing numbers (Ferrand de Almeida et al., 1983). Margarida Santos Reis has recently presented a composite distribution map for Portugal (Santos Reis, 1983). The otter is widespread throughout the country (Fig. 3.14), including parts of the rocky but sheltered coasts (Simões, 1977–1982).

In neighbouring Spain, Blas-Aritio (1978) studied otter distribution by means of questionnaires and suggested that both distribution and abundance had decreased markedly between 1966 and 1976. In 1981, Keith Elliot spent one month in Spain visiting sites over much of the country. He found signs at 70 (40%) of 176 sites and concluded that the species was most abundant in the uplands of the north and north-west and in the plains and hills of the south-west (Elliot, 1983). His results

Fig. 3.14. Distribution of otters in Portugal and Spain. Dots represent positive sites. (Data from Macdonald and Mason, 1982*b*; Elliot, 1983; with additional records from Santos Reis, 1983, and Delibes and Callejo, 1985.)

differed somewhat from those of Blas-Aritio, who did not note the importance of rivers like the Guadiana in the south-west nor the scarcity of otters in the Ebro basin in the north-east. Elliot cited water pollution as an important limiting factor in some regions.

Delibes and Callejo (1985) collected field observations between 1979 and 1983 and, while clearly they have more data than Elliot could gather in a month, they state that their results coincide with his rather than with those of Blas-Aritio. They, like Elliot, concluded that otters have declined in the plains and in intensively cultivated areas and are absent from rivers with high pollutant loads such as those in the Basque Country. Delibes and Callejo also suggest that otters are more common in regions with a rainfall greater than 600 mm per annum. They conclude that the species still exists in much of Spain (Fig. 3.14) but relate the marked decline over the last 20 years to the rapid economic development in the country since 1960. A detailed field survey was initiated in 1984 and provisional results indicate that 30% of sites were

Fig. 3.15. Distribution of otters in Italy and Yugoslavia. Dots represent positive sites. (Data from Macdonald and Mason, 1983a, and Liles and Jenkins, 1984, with additional records for Italy by Fabio Cassola (*in litt.*).)

positive (Miguel Delibes, pers. comm.). Delibes and Callejo are not very optimistic for the future of the otter in Spain, and economic growth in Portugal could also soon pose ecological problems there, but, at the moment, the Iberian Peninsula represents a valuable stronghold for the species.

In Italy, by contrast, the otter may well be on the verge of extinction. In 1975, Cagnolaro *et al.* published the results of a survey by questionnaire concluding that the species was widespread although depleted in numbers. Cassola (1980) doubted these findings and thought that the animal was probably rather rare. Tinelli and Tinelli (1980), noting that otters were common until the beginning of this century, blamed initial losses on extensive land drainage programmes encouraged during the Fascist Movement of 1922–1940 but they felt that a substantial decline had occurred since 1970.

In 1982, with the backing of the Italian World Wildlife Fund, we surveyed 188 sites in southern Italy. Signs of otters were found at 16

Fig. 3.16. A river in Calabria, southern Italy. The river is shallow and meanders across a broad, gravelly flood plain. The gravel is exploited on a large scale, with much disturbance and destruction of bankside scrub. Furthermore, many rivers are entrained by extensive gabions (large stones in steel cages), even in wild places, and bankside vegetation here is non-existent.

(8·5%) (Macdonald and Mason, 1983*a*). The distribution of positive sites suggested that small populations survived on the catchments of the Sele and Fiora, while some animals still lived on the Crati system, but signs in other regions were few and widely scattered (see Fig. 3.15). In southern Italy, the usual factors that adversely affect populations were present on a grand scale. Many rivers were grossly polluted by both domestic and industrial wastes. Bankside vegetation had often been removed or destroyed by regular cutting or during improvement of agricultural land. Widespread extraction of river gravel left banks devoid of cover and waters milky with suspended sediments (Fig. 3.16). Waterways were overfished and the level of hunting, with two million hunters in Italy (Cassola, 1979), makes security a scarce commodity for Italian wildlife.

Despite the problems, Italian conservationists are showing great concern for their otter populations. The enthusiastic Gruppo Lontra Italia is presently carrying out field investigations that, by the summer of 1985, should result in a complete coverage of Italy. A few previously unknown otter sites have already been located and Prigioni (1983) has confirmed their presence at the Lake of Mezzola in Lombardia. Yet, in

Fig. 3.17. Mikra Prespa, an internationally important wetland, looking from Greece across to the Albanian side. As well as a good otter population, the lake supports many waterfowl, including pelicans, that are endangered in Europe.

spite of these findings, the position of the otter in Italy still seems precarious.

Until very recently, information from Yugoslavia had been sparse. Reuther (1980*b*) suggested that otters probably occurred throughout the country but at a low density. There are six republics in Yugoslavia (two of which still permit otter hunting) but current internal information is only available from Slovenia in the north-west (Hönigsfeld and Adamič, 1983). In this region, where otters were widespread until 1960, records have been collected from hunting clubs. Since individual clubs operate in areas averaging only 40 sq km, individual otters may feature in more than one record and overall results may prove overoptimistic. However, accepting this flaw, these data are being used as the only available starting point for further investigations in the republic.

In the summer of 1982, Geoff Liles and Lyn Jenkins surveyed 129 sites in Yugoslavia and found signs of the animal at 57 (44%) (Liles and Jenkins, 1984). The best results derived from the catchments of the Drina and Bosna, although elsewhere populations appeared to be fragmented (Fig. 3.15). Liles and Jenkins considered the habitat to be suitable for the species, in terms of available cover, at 73% of sites visited, a situation reminiscent of Elliot's findings in Spain. It may be that in Yugoslavia water pollution is limiting populations, as suggested by Hönigsfeld and Adamič. Very large intensive pig units, for example, are causing severe pollution of some waterways (Marjana Hönigsfeld, pers. comm.).

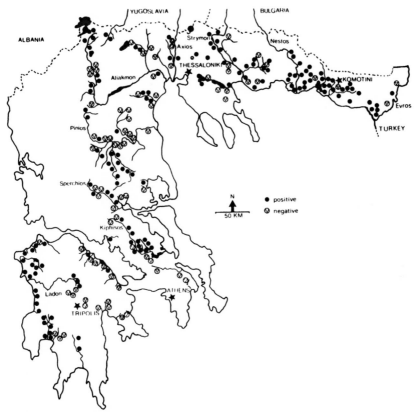

Fig. 3.18. Distribution of the otter in Greece (based on a field survey in 1981 (Macdonald and Mason, 1982a), with additional information collected in 1983 and 1984). The western mainland has not yet been surveyed.

To the south, Yugoslavia borders Albania and Greece. In Albania the situation of the otter is, and probably shall remain, a mystery (see p. 83), but it is just possible that viable populations occur since the animal is widespread in Greece and was found at Lake Mikra Prespa on the Albanian border (Fig. 3.17). The first field survey in Greece was carried out in 1981 (Macdonald and Mason, 1982a), when 62% of 200 sites produced signs (Fig. 3.18). Otters seemed to be common in the western Peloponnese, in the lowlands of the Kiphisos catchment, in the Nestos delta and east of Kavala towards the Bulgarian and Turkish borders. It was felt that populations in the eastern Peloponnese and in the lowlands west of Thessaloniki were fragmenting and further work in 1983 (Macdonald and Mason, 1985) seemed to confirm that rivers like the Axios and Aliakmon were less suitable for the species. Otter signs were found more frequently on waterways offering dense bankside vegetation,

Fig. 3.19. (*a*) This upland river in north-east Greece is lined by *Alnus* and *Salix*, beneath which tangles of *Rubus* cover fallen rock. Dense maquis extends from the hillside to the river's edge and in such habitat can be found an abundance of safe refuges. (*b*) An irrigation ditch in the Nestos delta, where dense growths of *Phragmites* and *Typha* can shelter otters. In Greece, such ditches are still widely used by the animals, but for how long will this situation last?

with *Rubus* being especially important on upland rivers, dense *Salix* scrub on both upland and lowland rivers and *Phragmites* in irrigation channels (Fig. 3.19). However, the rivers Axios and Aliakmon, as well as providing poorer cover, may also carry pollutants from Yugoslavia. On the Evros catchment too, where otter distribution seemed rather patchy, pollution from Bulgaria and Turkey may be having an effect.

It is quite unusual now in western Europe to find thriving otter populations in irrigation ditches surrounded by flat fields of maize. At present in Greece, because vegetation in the channels is cut on rotation and since regrowth is rapid, the ditches provide ideal cover for otters. However, increased herbicide use could change this and, as in Portugal, more pesticides may also be applied in the future. We felt, in 1983, that a slight otter decline may have begun in the Nestos delta and it may be significant that residues of organochlorine pesticides were detected in spraints we collected from this region in 1984. By contrast, no pesticides were found in spraints from the uplands of the River Kuru. Typically, as otter populations decline, they disappear initially from agricultural lowlands.

Distribution in eastern Europe

In eastern Europe some data on otter distribution have derived from questionnaire returns, but generally information is sparse and not always convincing. Reuther (1980*a*), from what was available, concluded that the species was probably widespread in both Bulgaria and Romania, although in Romania numbers appeared to be generally low, with clear signs of a decline in some areas. In Bulgaria, stable populations could be found, particularly in the south, but the animal was rare in mountainous regions. Spiridonov (1985) thought that, since the turn of the century, otter numbers in Bulgaria had decreased six-to eight-fold until, in the past 20 years, there had been some recovery. He thought that the rapid development of fish farming had benefited the species and considered that, today, good populations could be found in the Balkans and on the coast.

In Hungary, Tánko and Tassi (1978) used Ministry of Agriculture and Food shooting statistics to demonstrate that otters were rare in the east, south-east and north. In the west and south-west, however, viable numbers still occurred. Nechay (1980) confirmed that the greatest density of otters could be found in the west and suggested that, since

1973, declines had occurred east of the Danube. He thought that, as in Bulgaria, the species had benefited from the growth of fish-farming, but then most of his information came via complaints of damage from fish farmers. Kiss (1985), using some of her own field observations, suggested that, in the face of reported economic damage by otters, Hungarian conservationists, in some regions, now had to deal with problems of keeping the population at 'an optimal level'.

For Czechoslovakia, Anděra and Trpák (1981) give distribution maps for otters in 1920–1930 and 1970–1975. Between these two periods, declines have taken place except in the south and south-west. Baruš and Zejda (1981), having analysed returns of questionnaires sent out in 1978, calculated a minimum of 174 otters in the Czechoslovak Republic and considered the best area for otters to be the South Bohemian region, where waters were unpolluted and numerous fish ponds were connected by streams. Kučera (1980) considered that 62% of all waterways in Czechoslovakia still offered suitable conditions for otters and in the region where he worked, on the borders of the Bavarian and Bohemian forests, he reported a stable population of 25 animals. Hell (1980) calculated from 1977 hunting statistics 208 otters in Slovakia and 73 for Bohemia and Moravia. This total had risen to 355 by 1978. Whether such precise figures are in any way meaningful is open to doubt. Novakova (1985) suggested to the Strasbourg Otter Colloquium that, while the situation of the otter in Czechoslovakia was not too encouraging there was, at present, no need for concern. From the outside, the picture is confusing, as is so often the case with information derived from questionnaires.

In East Germany, Stubbe (1977), having appealed for information through journals and by means of 7000 questionnaires, concluded that only three regions still provided conditions suitable for otter reproduction. These were the Lausitz region east of the River Elbe, the southern part of the Frankfurt/Oder area and the lakes of Mechlenburg. He estimated the G.D.R. otter population at 600 ± 200 individuals, the same estimate he gave in 1984 (Stubbe, 1985a).

In neighbouring Poland, Pielowski (1980) suggested that otters occurred throughout the country, but at a low density. He considered the best regions to be the swamplands of the east and the mountain rivers of the Beskids and Carpathians. Wlodek (1980) studied the species in West Pommerania in north-west Poland. From a survey made between 1974 and 1979 he estimated 296 ± 100 animals in the area but thought numbers were declining. Later he suggested (Wlodek, 1985) that poachers and musk rat traps were major causes of otter mortality. Sikora (1984, 1985) used circulated questionnaire returns to assess otter distribution since 1977 and concluded that the species was rare and in decline. He collected reports of the animal from 110 localities and found numbers to be decreasing at 30 of these.

Novikov (1956) described the otter as widespread in Russia although numbers were generally low and in some regions it had been exterminated. Animals could still be found from the Barents Sea, including some offshore islands, to the Caucasus in the south and through to the Far East. In some parts of the south it was said to be common. Heptner and Naumov (1974) reported that otters were not numerous in Russia, and Kucherenko (1976), studying the species in the Amur-Ussuri District, thought that the estimated 12000–14000 animals present in 1966–1968 had subsequently declined four- to five-fold.

Distribution in Asia

To the south-west, Russia borders Turkey, a country where viable otter numbers may well remain but, as with so many eastern countries, information is negligible. Müffling (1977), having asked for information from the Ministry for Hunting and Fishing, was informed that the species was widely distributed but in low and declining numbers. The most suitable areas for otters were considered to be the eastern Mediterranean region and along the Black Sea. Reuther (1980b) mentioned the north-east and also the central Taurus as places where thriving populations might be found but, in the regions of the Bosporus and Marmora Sea, otters seemed to be rare.

Information of this type, all that can be gleaned, means little and it would be of interest to know something of the true status of otters in Turkey.

Within the Mediterranean Basin, we now know that otters thrive in parts of Iberia, in Greece and in some regions of North Africa (see p. 78) but in the eastern Mediterranean we are faced, yet again, with limited information. According to Benny Shalmon (in litt.) the Israeli otter population is diminishing. In 1970–1976 the number of animals was estimated at between 40 and 70, with most of these in the Jordan river systems. The species still occurred also in the coastal plains between Haifa and Acre, and in rivers south of Haifa. The current status of the otter in Israel is unknown.

For the remainder of the Middle East, Harrison (1968), based on dated information, recorded otters in Jordan, Lebanon and Iraq, but there are no data on status.

From countries further east, information is virtually non-existent. The East German, Stubbe, recently visited Mongolia and reported his findings to the Strasbourg colloquium. He considered otter distribution in Mongolia as a mosaic with some good populations in the north and with animals still to be found in the east and west of the country. In central regions, the species seemed to be rare and, not too surprisingly, none was found in the Gobi. Overall, Stubbe (1985*b*) found no evidence of population increases.

In neighbouring China, otters are now considered rare having suffered reductions in both numbers and range (Ma Yi-ching, *in litt.*). In the Mai Po Marshes, the last extensive wetland in Hong Kong, the otter was common up to the 1940s but is now extinct (Melville and Horton, 1983).

Wayre (1978) visited Sri Lanka in 1975 and found the otter to be widely distributed from sea level up to at least 1800 m. It seemed to be common in paddy fields and in streams with adequate bankside cover. Chitampalli (1979) reported finding *Lutra lutra* at lakes in Bhandara District (north-east Maharashtra) in India; previously it had been thought to occur only in the south and in Kashmir. The otter also occurs throughout South-east Asia (Fig. 2.1, p. 8), but there is no information on status.

At the eastern edge of its range, in Japan, Mikuriya (1976) suggests that the otter, which he calls *L. l. whiteleyi*, is on the verge of extinction.

Distribution in North Africa

Lutra lutra also occurs in North Africa, specimens from Morocco once being described as *L. l. splendida* because of their cinnamon colour and unusual dentition (Cabrera, 1932). In 1983, we visited the region of Morocco lying to the west and north of the Haut Atlas and surveyed 78 sites. Signs of otters were found at 46% (Macdonald and Mason, 1984). In lowland areas, the animal is scarce but can still be found along the Oued Tensift and Oued Oum-er-Rbia and at the tidal lagoon of Merja Zerga, a wetland recognized by the Ramsar Convention. However, 75% of the sites where we located otters were in foothill country. Our records, together with those collected by Broyer *et al.* (1984), are shown in Fig. 3.20. In the region we visited, otter distribution was limited by

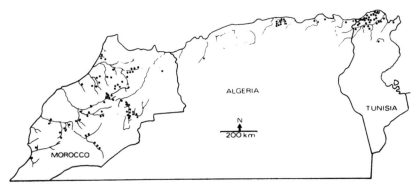

Fig. 3.20. Distribution of the otter in North Africa. Dots represent positive sites. (Data from Macdonald and Mason, 1983*b*, 1984, 1985, and Broyer *et al.*, 1984, with additional records for Algeria by Koen de Smet (*in litt.*).) The extreme east and extreme west of Algeria await a field survey.

a lack of water, since there are few rivers between Casablanca and Agadir or inland to Marrakesh. Many dams have been constructed in the last 30 years (Abouzaid and Hajji, 1982) and their outflow catchments are frequently dry. In addition, water is often supplied to the arable plains in raised concrete channels rather than in irrigation ditches such as those used by otters in Greece. Bankside habitat was more often suitable in the uplands where tangles of *Nerium oleander*, *Rubus* and *Rosa* provided secure shelter. Upland habitats also provide shelter amongst rocks (Fig. 3.21). Many lowland sites were rather bleak. There is little evidence of water pollution in Morocco despite rivers being heavily used for domestic and agricultural purposes. Waters in the arable plains may, of course, be contaminated with agricultural chemicals.

In Tunisia also, rivers appear to be free of pollution and the otter is common in the extreme north of the country (see Fig. 3.20). In 1982, 75 sites were surveyed and signs were found at 30 (40%). Within the region to the west of Tunis and north of the Oued Medjerda, 71% of sites proved positive. The catchment of Lake Ichkeul appeared to be of particular importance to the species but, now that dams are being constructed on the inflow rivers, resultant changes in flow rates and increased salinity in the lake could prove inimical. Within this northern region of Tunisia, signs could be found even on tiny streams and at scattered pools, suggestive of a healthy population utilizing all available habitat (Macdonald and Mason, 1983*b*).

The distribution of the otter in north Tunisia appeared to be coincident with the availability of dense riparian cover provided mainly

Fig. 3.21. At first sight, this semi-desert habitat east of Agadir, Morocco, looks unpromising for otters. However, pools in the small stream held abundant cyprinids and cavities amongst the rocks provided shelter for the animals. Many spraints were found along this stretch of water.

Fig. 3.22. In north Tunisia, otter distribution appears to be related to the availability of bankside cover provided by *Nerium oleander*. Since it is poisonous, this plant is not overgrazed by sheep and goats and offers impenetrable year-round cover.

Fig. 3.23. In Grande Kabylie a few otters still remain, but here too some water pollution, typical of much of Algeria, was evident. At this site the habitat was ideal for otters, comprising scrub growing amongst loose rocks. The stream was, however, grossly polluted with effluent from villages in the hills.

by *N. oleander* (Fig. 3.22). To the south of the Oued Medjerda, where rivers run through semi-desert, often disappearing underground, there was little or no form of shelter.

With otters still thriving in the Moyen Atlas of Morocco and in the north of Tunisia, it might be supposed that Algeria too would support viable numbers. Sadly, this does not appear to be the case. In 1984, we visited 52 sites in north-central Algeria and found signs of otter at only 10 (see Fig. 3.20). All the positive sites were in a limited area within the catchment of the Oued Sebaou. Throughout the area studied, gross organic and industrial pollution of waters was widespread, and some rivers were lifeless due to heavy metal contamination from mines. Pesticides and herbicides are widely used but without instructions on methods of application. Many of the rivers we visited offered little cover for otters and food supplies appeared to be limited by pollution or by seasonal desiccation. Indeed, in Algeria, one can find all the environmental problems so common in parts of Europe and the gross pollution of waters contrasts sharply with the situation in neighbouring Morocco and Tunisia.

In our visit to Algeria, we could only find otters in the east, in Grande Kabylie, where the hills are still wooded and where fewer rivers are, as

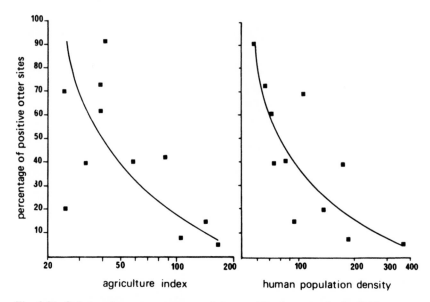

Fig. 3.24. Relations between percentage of sites positive for otters during field surveys (y) and (a) an index of agriculture (x_1 = total crop production/total land area, $m_t \times 10^3/(km^2 \times 10^3)$) and ($b$) human population density (x_2, people/km^2) for 11 countries. Production and population density are for the year 1978. The relationships can be described by the equation arcsin $y = 186\cdot4 - 16\cdot6 \ln (x_1, x_2)$; $r^2 = 0\cdot88$.

yet, polluted (see Fig. 3.23). But, even here, the level of marking by the animals suggested that the population was low. It may well be that further east, towards the Tunisian border, the habitat is more suitable for the species and otters may also thrive on the Moroccan border and south of the Atlas mountains. However, within the north-central region, the otter is certainly endangered. Its decline can be linked to the rapid increase in human population and recent advancements in industry and agriculture lacking any coherent environmental planning. The Algerian government's present policy of industrial de-centralization can only limit otter distribution still further.

Conclusion

This chapter has described a widespread decline in otter populations over much of the species' range. The activities of man have been responsible for these population decreases. Figure 3.24 shows relationships between the results of field surveys for otters and two rough indices of human pressure on the environment – population density and crop productivity. The factors influencing otter survival are described in the next chapter.

Note added in proof: Claudio Prigioni visited Albania in May 1985 and recorded otters at 16 out of 31 sites examined.

4 FACTORS AFFECTING OTTER SURVIVAL

It levelled every bush and tree and levelled
every hill
And hung the moles for traitors – though
the brook
is running still
It runs a naked stream, cold and chill

JOHN CLARE,
Remembrances

The previous chapter has catalogued the declines in populations that have occurred over much of the otter's range. In many regions the otter is now extinct, or is on the verge of becoming so. In Britain, from where we have most information, there are rivers from which the otter is still retreating. In this chapter, we examine the factors behind the decline. At the outset it must be emphasized that no single factor can be identified as the sole cause of the decline and at present a number of adverse influences may be militating against the return of the otter.

Pollution

For the present purposes, the definition of pollution given by Mason (1981) can be modified to read 'the introduction, by man, into the environment, of substances liable to harm otters or damage their resources'. Pollutants may therefore affect otters indirectly, through damaging their food supply or habitat, or by exerting direct toxic effects. Of course, some pollutants may simultaneously kill both otters and their life support system.

Indirect effects

The most important pollutants exerting indirect effects are those that destroy fish populations. The most widespread of these pollutants are biodegradable organics, particularly sewage, which decompose within the waterway, utilizing the available oxygen, and industrial effluents, containing such chemicals as chlorine, ammonia and phenol, to which fish are highly susceptible.

In England and Wales, a national survey of rivers during 1980 recorded 610 km of rivers and canals that were grossly polluted and a further 3010 km that were polluted to such an extent that fish were absent or only sporadically present (National Water Council, 1981). This represented 7% of the total length of 38 740 km. In the same survey, it was reported that 16% of 2800 km of the lower tidal reaches of rivers were of poor quality or grossly polluted. However, vitually all of these polluted, fishless reaches of rivers and canals lie within the urban conurbations of South Wales, the West Midlands, South Yorkshire, Tyneside, Greater Manchester and Merseyside, which are also highly disturbed and lack suitable habitat. In these areas, otters would undoubtedly be absent irrespective of any improvement in water quality. Nevertheless, occasional pollution incidents may pose local problems to otters (Fig. 4.1).

An increasing problem in Britain, highlighted in the National Water Council report, is of organic pollution from farm wastes. This is the result of the intensification of animal production, with attendant problems of waste disposal, and is undoubtedly exacerbated by a reduced level of monitoring and policing by the water authorities, due to a cut-back of staff during the economic recession. The problem seems most acute in south-west England and is affecting previously clean rivers with good otter populations (Elizabeth Lenton, pers. comm.). The situation needs watching but, provided discharges are intermittent and small-scale, reductions in oxygen may be insufficient to affect fish populations. We well remember that, after a day and night of unseasonable heavy rain in Portugal, the river below Portalegra was polluted with raw sewage from the overloaded sewage works. Pond turtles were obviously under stress, their heads protruding out of the black water, but we saw no dead fish. They may have sought refuge in deep pools while the plug of filth passed. Abundant signs of otter were found.

In Britain, there has been a long history of pollution control, but in other European countries, long stretches of rivers running through rural areas may be fishless due to pollution. For example, in Spain, Elliot

Fig. 4.1. (*a*) Organic effluent from a food factory, situated in rural England, was causing gross pollution of this otter river in January 1984. The primitive treatment facilities at the factory were severely overloaded. Sewage fungus (*b*), a complex community of bacteria and protozoa, typical of organic pollution, thickly coated tree roots, aquatic vegetation and the river bed, smothering the natural river bed community for a distance of 20 km downstream. The pollution inspector of the regional water authority, when informed of the situation, acted quickly to prevent further discharge and the river cleaned within a few days. Because of high winter flow in the river, the pollution did no permanent damage. Had it occurred in summer, a massive fish kill might have resulted. The company have now installed an efficient treatment plant at this factory. Constant vigilance is required on the part of all river users to detect such pollution incidents. The recent recession has resulted in a severe cut-back in the staff and facilities of water authorities, such that these organizations appear no longer to be able to respond adequately to pollution incidents – this pollution should have been detected early by a water authority inspector, not belatedly by an otter biologist.

(1983) reported the River Tajo and its tributaries as grossly polluted and fishless below Madrid, for a distance of some 300 km. After passing through several reservoirs the water purifies and, by the time the Tajo flows across the border into Portugal, otters are present. Elliot considered that industrial pollution was restricting the distribution of the otter in certain areas of Spain, particularly the north-east. Green and Green (1981) considered that widespread organic and industrial pollution were limiting the distribution of otters in France, while we have noticed extensive stretches of polluted waterway, with no signs of otters, in Belgium. It would appear then, that, at least in parts of mainland Europe, the discharge of organic and industrial effluents into rivers may, by limiting fish populations, render otherwise suitable habitat unsuitable for otters. Nevertheless, we know no definite evidence where a pollution incident has destroyed a fish population and resulted in the desertion of an area, even temporarily, by otters.

Over the last decade, acid rain has emerged as a major pollution problem. Sulphur dioxide (SO_2) and nitrogen oxides (NO_x), produced while burning fossil fuels, are transported long distances in the atmosphere, to be deposited as sulphuric and nitric acids on land and in water, rain, snow or dust. Tall smoke stacks built at electricity generating stations reduce local pollution, but the waste gases are caught up by the prevailing winds. Pollution from Britain and central Europe falls as acid rain in Scandinavia, while pollutants from industrialized regions of the United States are transported to Canada. Acidified rain has also been recently identified in northern Britain (Fowler et al., 1982).

Acid rain causes tremendous damage to coniferous forests, but here we are concerned with its effects on streams and lakes. In water already low in nutrients (oligotrophic waters with a low buffering capacity), the acidity increases to such an extent that fish and invertebrates decline in numbers and eventually disappear. Acid rain also dissolves toxic metals from rocks and soils, which then enter watercourses; aluminium is particularly toxic to fish. In Sweden, some 20 000 lakes are now lifeless or severely affected by acidification (Chadwick, 1983; Swedish Ministry of Agriculture, 1983). With a loss of fish over wide areas, we can predict that otters will disappear. However, there have been no studies on changes in otter populations during the progress of acidification. Nevertheless, Olsson and Sandegren (1985) found five populations of otters, all on well-buffered eutrophic waters in the counties of Södermanland and Småland in southern Sweden; no otters were found on oligotrophic waters.

Another widespread problem affecting freshwaters is that of eutrophication (Mason, 1981), though this pollution could be beneficial to otters. Eutrophication is the addition to rivers and lakes of plant growth nutrients, especially nitrogen (mainly from fertilizers) and phosphorus (chiefly from treated sewage effluents). The resultant increase in plant productivity also results in an increase in fish biomass, with salmonids being replaced by cyprinids (Hartmann, 1977), a preferred food of otters. As described above, eutrophic waters are incidentally buffered against acidification. Again, otters have not been studied in relation to eutrophication.

High concentrations of nitrates in rivers are considered to be human health hazards, when the water is abstracted for potable supply. Nitrate levels have been steadily increasing in rivers in eastern England over the past few decades, as increased amounts of fertilizer are applied to the land, and many rivers now have nitrate concentrations above World Health Organization standards, at least for part of the year. Young babies are especially at risk, while there is some evidence, but by no means conclusive, that nitrates are converted to carcinogenic methylamines in the body. Whether otters run similar risks is unknown.

Direct effects
Thriving otter populations exist on northern coasts (p. 60) and fears have been expressed about the possible effects of offshore oil developments on them. Oil spills may also occur on rivers. In July 1982, we observed the main river, the Medjerda, in Tunisia running black with oil from an accidental spill; there were otters living downstream.

An oil spill occurred at Shetland's Sullom Voe Oil Terminal in December 1978, when 1200 tons of bunker C fuel oil were spilled into the sea, resulting in the deaths of at least 13 otters (Baker *et al.*, 1981). The otters did not appear to recognize oil as a danger and became heavily contaminated. Post-mortems on a number of these victims revealed that they had ingested large amounts of oil. Blood was present in their stomachs and it was concluded that they died of haemorrhagic gastroenteropathy.

The sea otter *Enhydra lutris* is also potentially vulnerable to oil pollution. Apparently over 100 were killed by a spill of diesel oil in the U.S.S.R. (Van Blaricom and Jameson, 1982). In an experiment, two sea otters were observed in a swimming pool that was partially contaminated by oil (Siniff *et al.*, 1982). Both otters appeared to avoid the oil, but they occasionally surfaced in it and became contaminated. One died within a day of becoming oiled. Otters were lightly oiled and released at sea with radio-transmitters attached. They were noted to increase their level of activity, chiefly because they spent longer grooming. Sea otter fur, when contaminated with oil, has an increased thermal conductivity (Kooyman, Davies and Castellini, 1977), so that oiled sea otters suffer from increased heat loss, which they must counteract by raising their metabolic activity. It is likely that the Shetland otters ingested oil during grooming to clean their fur.

Provided that oil spills remain irregular and local, they are unlikely to cause long term declines in otter populations. Accidental oil spills will never be eliminated. Long experience with sea birds, which have suffered massive mortalities due to oil pollution, suggests that, although the sight of oiled birds causes deep offence to bird lovers and much suffering to individual birds, populations are not generally being damaged (Clark, 1984).

Of more widespread concern are those toxic chemicals that are dispersed widely in the ecosystem, are persistent and are stored in tissues. General reviews of toxicity and wildlife are presented by McEwen and Stephenson (1979) and Bunyan and Stanley (1982). Pollutants of particular concern in relation to otters are chlorinated hydrocarbon pesticides, polychlorinated biphenyls and heavy metals. Before describing the evidence for their accumulation in otters, it is first necessary to discuss the problems of interpreting ecotoxicological data.

Evidence for the accumulation of toxic chemicals in otters comes mainly from the analysis of tissues. As otters are both rare and protected over most of their range, tissues are normally obtained from animals killed accidentally, for example those hit by vehicles or drowned in fish traps. It is assumed that such animals are representative of the population as a whole, though it is entirely feasible that individuals with higher pollutant loads may be more susceptible to accidents. Because few of the animals killed in this way are reported to the authorities, data build up very slowly. In North America, where the Canadian otter (*Lutra canadensis*) is still hunted for it pelt, tissues are more readily available, though surprisingly little use has been made of them.

The tissue concentrations of pollutants that are likely to cause harm to otters are unknown. If observed levels are high, we may be justified in extrapolating from information on other species, e.g. domesticated animals, which have been more fully studied. However, because individual species are highly variable in their response to toxic chemicals, it is extremely difficult to assess the importance of lower concentrations. It is these lower levels, which may exert sublethal effects, that could be of significance to an otter population in the long term.

The scientific ideal would be to conduct laboratory experiments, dosing otters with a range of concentrations of a particular pollutant and studying the effects at various levels of organization, e.g. biochemical, physiological, behavioural. At the end of the experiment the otters would be killed, the concentrations of the pollutant in the tissues would be measured and a pathologist would screen for tissue abnormalities. Hopefully, the ecotoxicologist could then state, with some confidence, that particular concentrations of a pollutant in the environment would produce particular adverse effects on wild populations of otters.

Because of the otter's rarity, most people would consider such experiments unethical. Because of its specialized requirements, the otter is a highly unsuitable animal for experimentation. We know of only one such attempt (p. 99). Such laboratory experiments are also often meaningless in environmental terms. Normally, toxicological tests are carried out at concentrations in excess, and often greatly in excess, of concentrations encountered in the environment so that unambiguous results can be obtained. The stresses faced by a well-fed animal in a laboratory situation bear little relationship to those faced by a wild animal, seeking food, avoiding predation and breeding in an environment that is frequently inclement. Furthermore, the critical stage in the life cycle may not be the adult; embryos and foetuses are often more susceptible to pollutants. Finally, the wild otter may be carrying a load of several pollutants, all of which may be potentially harmful. It is not known whether the effects are additive.

The biologist must, then, make a subjective assessment of the significance of the pollution loads he records in otter tissues. Bearing in mind the severe declines that have occurred in otter populations, the prudent conservationist may express concern over *any* pollutant found in otters. Other scientists, for example those employed in the pesticide industry, may subjectively interpret the data in an entirely different way.

Persistent pollutants, which accumulate in living tissues, are a particular problem in fresh waters because there are many sources. Rainfall will wash atmospheric pollutants and chemicals applied to land into watercourses. Some pesticides are deliberately applied to water, for instance to control mosquitoes and black-flies. Many industries discharge effluents directly into rivers, or indirectly via sewage works. Very small amounts of persistent pollutants in effluents may become quickly concentrated in the biota. Fish, for example, will take in pollutants with their food, via their gills and through their skin. Otters obtain their pollutant load almost entirely from their food but, if this is contaminated fish, heavy doses are quickly accumulated.

Of pesticides, the chlorinated hydrocarbons give rise to most concern. As well as being used in agriculture, horticulture and forestry, they are used as moth-proofers in the carpet-making and woollen industries and as wood preservatives. They dissolve readily in animal fats and hence accumulate in tissues. When these fats are mobilized during periods of stress, such as food shortage or reproduction, large amounts of pesticide may be released into the blood stream, with extremely toxic results. Organochlorine pesticides induce the hepatic microsomal enzyme system, which is involved in the synthesis of many compounds within the body.

DDT was first recognized as an insecticide in 1939 and by the early 1950s was being widely used in agriculture. In body tissues it is metabolized to DDE. Lindane (gamma BHC or HCH) was introduced at about the same time. During the 1950s the cyclodiene chlorinated hydrocarbon insecticides were introduced, compounds such as aldrin, dieldrin and heptachlor. The active ingredient in dieldrin is known as HEOD and aldrin also breaks down to HEOD in the environment and in tissues. The cyclodienes are more toxic to mammals than DDT. During 1954 and 1955, 1580 ha of land in Illinois, U.S.A., were treated with dieldrin to destroy Japanese beetle and populations of cottontail rabbit, muskrats and ground squirrels were virtually exterminated (Scott, Willis and Ellis, 1959). Many foxes in Britain died during the late 1950s (Turtle et al., 1963) and death was brought about by eating pigeons, which themselves had been feeding on treated grain. A tissue level of 1 p.p.m. (part per million or mg/kg) was thought to be diagnostic of dieldrin poisoning in foxes (Blackmore, 1963). In contrast to its toxicity, DDT is much more persistent in the environment than dieldrin. It has recently been estimated that the half-life of DDT in uncultivated soils was 57 years (Cooke and Stringer, 1982), compared with a half-life of dieldrin of from four to seven years (Edwards, 1966). There is, then, a source of contamination for many years to come. Chlorinated hydrocarbons are so widely disseminated in the environment that, in Britain, none of 500 sparrowhawks (*Accipiter nisus*) or their eggs analysed was free of residues (Newton and Haas, 1984).

On the basis of records of the hunts the otter in Britain declined sharply in numbers during the late 1950s and 1960s, (p. 57). Chanin and Jefferies (1978) attributed the decline to the introduction of a new

Table 4.1. *Concentration (p.p.m.) of dieldrin and DDT in otter tissues*

	Locality	Number of samples	Tissue	Dieldrin		DDE		Total DDT		Authority
				Mean	Range	Mean	Range	Mean	Range	
Lutra lutra	Great Britain	32	Liver, wet weight	0·5	0–13·89					Chanin and Jefferies, 1978
	Great Britain	14	Muscle, extractable fat	6·5	0–29·0			18·3	0–85·0	Author's unpublished data
	Norway	23	Muscle, extractable fat					1·7	0·18–5·9	Olsson, Reutergårdh and Sandegren, 1981
	Sweden	53	Muscle, extractable fat					4·1	0–27·0	Olsson et al., 1981
Lutra canadensis	U.S.A.									
	Oregon	20	Liver, wet weight					0·89	0–4·95	Henny et al., 1981
	Oregon	20	Leg muscle, wet weight					0·31	0–1·5	Henny et al., 1981
	Alabama	13	Muscle, wet weight	0·006	0–0·04			0·96	0·7–7·49	Hill and Lovett, 1975
	Georgia	102	Adipose tissue			8·72	0·2–136·7			Halbrook et al., 1981
	Georgia	9	Adipose tissue	0·33	0·02–2·0					Halbrook et al., 1981
Enhydra lutris	California	10	Adipose tissue					11·1	0·1–36·0	Shaw, 1971

Note: Concentrations expressed in terms of fat will be higher than wet weight values.

pollutant into the environment and suggested dieldrin as the cause. A sample of 32 otters has been analysed and 26 had measurable quantities of dieldrin, with a mean level in the liver of 0·5 p.p.m. One animal had a liver concentration of 13·95 p.p.m. dieldrin, which was almost certainly lethal (Jefferies, French and Stebbings, 1974). The decline of the otter coincided with those of other predatory species, for which there is more information. For example, the main decline in sparrowhawk numbers was between 1957 and 1963 and levels of dieldrin were greater in sparrowhawks from areas in which the decline had been most severe (Newton and Haas, 1984). Details of the analyses of a series of otter tissues from Britain have unfortunately not yet been published, so that a full assessment is not possible.

A summary of the data on dieldrin and DDT in otter tissues is given in Table 4.1. Small quantities of other pesticides, such as lindane, mirex and chlordane, have been found in limited areas. If a tissue concentration of 1 p.p.m. causes toxicity in otters, as it appeared to do in foxes (Blackmore, 1963), then otters in Britain and Georgia could have died of dieldrin poisoning, while the mean levels of dieldrin might have been high enough to exert sublethal effects, for instance on reproduction. The levels of DDT, though higher, are unlikely to be having adverse effects. It must be remembered that these analyses were carried out on animals collected in the 1970s and 1980s, whereas the main usage of chloro-hydrocarbons was in the late 1950s and early 1960s, so that much higher tissue concentrations may have passed unrecorded. Our current knowledge of pesticide residues in otters is highly inadequate.

In Britain, because of the detrimental effects of chlorinated hydro-carbon pesticides on wildlife, a series of voluntary agreements have been introduced to curtail their use. The cyclodienes were voluntarily withdrawn from use on spring cereals in 1962, from use in sheep-dips in 1965 and from use on autumn-sown cereals in 1975. In 1981, further restrictions, but again voluntary, were imposed throughout the E.E.C. to phase out the uses of these pesticides. The decline in use of DDT, dieldrin and aldrin is shown in Fig. 4.2. Heptachlor, another cyclodiene, was applied extensively to crops until 1966, but has since been used in insignificant amounts. DDT is still widely used in agriculture, but there has been a decline in the application of aldrin and dieldrin to crops in the latter half of the 1970s.

With this decrease in usage of pesticides, predatory birds such as the sparrowhawk and peregrine (*Falco peregrinus*) began to increase in population (Newton and Haas, 1984; Ratcliffe, 1980). Some increase in

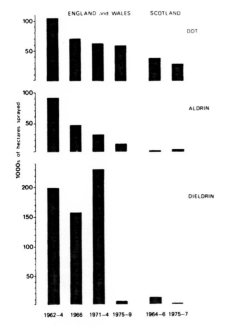

Fig. 4.2. Estimated annual usage of organochlorine pesticides in British agriculture and horticulture (thousands of hectares sprayed). Those areas with two separate applications of pesticide in a year would be recorded twice. Other usages of pesticides, e.g. sheep dip, moth-proofing, are not included. The information is based on sample surveys of farmers, providing information voluntarily. There is serious confusion, even among some officials, as to whether bans on pesticide usage are mandatory or voluntary (they are the latter in Britain), so that farmers may have been discreet over their real applications of pesticides, leading to underestimates in the official figures. (Data from Strickland, 1966; Wilson, 1969; Sly, 1977, 1981; Cutler, 1981.)

the former began within a few years of the first restriction on cyclodienes (1962). The recovery has occurred over all but those counties with the highest proportion of arable land, and has resulted from an increased survival of adults rather than an increase in breeding success, which has remained poor, due to DDE residues. Because cyclodiene levels have decreased in sparrowhawks throughout the 1960s and 1970s, whereas DDE residues have remained constant, Newton and Haas (1984) argue that cyclodienes have been largely responsible for causing population changes.

These improvements in the fortunes of birds of prey have not been matched with an increase in otter populations. Indeed the species is still declining (p. 61). Measurable quantities of dieldrin were found in tissues of 11 out of 16 otters found dead in 1984 (authors' unpublished data), though levels were generally low. We must therefore look for factors other than cyclodienes to explain fully the demise of the otter.

The second group of potentially harmful pollutants are the polychlorinated biphenyls (PCBs), which have many of the properties of organochlorine pesticides, being insoluble in water but highly soluble in animal fats, and very persistent in the environment. PCBs are not used in agriculture, but they have been incorporated into a wide range of industrial products, such as transformers, rubber, plastics and brake

Table 4.2. *Concentration (p.p.m.) of PCBs in otter tissues*

	Locality	Number of samples	Tissue	Mean	Range	
Lutra lutra	Norway	23	Extractable muscle fat	17·0	1·6–30·0	Sandegren, Olsson and Reutergårdh (1980)
	Sweden	53	Extractable muscle fat	120·0	4·7–970·0	Sandegren *et al.* (1980)
	Great Britain	14	Extractable muscle fat	53·0	0–300·0	Authors' unpublished data
Lutra canadensis	Oregon	20	Liver, wet weight	4·51	0–23·0	Henny *et al.* (1981)
		20	Leg musle, wet weight	1·84	0–8·3	Henny *et al.* (1981)
	Alabama	19	Muscle, wet weight	0·36	0–2·5	Hill and Lovett (1975)
	Georgia	55	Adipose tissue	8·24	0·6–66·7	Halbrook *et al.* (1981)

Note: Concentrations expressed in terms of fat will be higher than those expressed as wet weights.

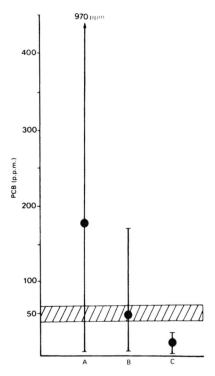

Fig. 4.3. Concentrations (means and ranges) of PCBs in extractable fat from muscle of otters from (*A*) southern Sweden; (*B*) northern Sweden; (*C*) Norway. The hatched area indicates the level of PCBs in the muscle of experimental female mink exhibiting reproductive failure. (Adapted from Olsson *et al.*, 1981.)

linings. The burning of waste plastics, containing PCBs as plasticizers, results in the discharge of PCBs to the atmosphere, where they may be transported over great distances. PCBs contain highly toxic impurities, such as dioxins and furans. They are a mixture of many compounds, so that on chromatograms they appear as multiple peaks, making quantification difficult and presenting problems in comparing the results from different laboratories. PCBs were not identified as environmental contaminants until 1966 (Jensen, 1972), though they had been used in industry for at least 35 years prior to that, and have since been proved to be widespread in the ecosystem.

The concentrations of PCBs recorded in otter tissues are shown in Table 4.2. In Sweden, the decline in the otter has been attributed to PCBs. The concentration of PCBs, in the extractable fat of muscle tissue of Swedish otters was much higher than in similar samples from Norway (Fig. 4.3), where populations are stable or increasing. DDT levels were comparatively low, but cyclodienes were not looked for.

Mink, another fish-eating mustelid, are bred for their fur and hence are more amenable to experimentation and can be used, with caution,

Table 4.3. *Effects of PCBs, fed for 66 consecutive days, on the reproduction of mink* (Mustela vison). *Data from Jensen et al.* (1977)

| | Group | | |
	A	B	C
No. of females	25	22	22
PCBs (p.p.m.) in diet	0·5	3·3	11·0
PCBs (p.p.m.) in extractable fat	14·0	86·0	280·0
Number of pups born/pregnant female	5·1	2·9	0

as models for otters. During the 1960s, ranch mink, fed on lake fish in North America, exhibited reproductive failures which were attributed to DDT, though other contaminants were not determined. Subsequent experiments in North America and Sweden have shown that a diet containing up to 100 p.p.m. DDT had no effect on reproduction in mink (Aulerich and Ringer, 1970; Jensen *et al.*, 1977). By contrast, PCBs fed at a low dose of 2 p.p.m. over eight months had a detrimental effect on mink reproduction (Aulerich and Ringer, 1977), while diets containing 10 to 12 p.p.m. PCB resulted in complete reproductive failure (Aulerich and Ringer, 1977; Jensen *et al.*, 1977). Some of the results of the Swedish experiments are given in Table 4.3. The females were mated to untreated males. Increased dietary PCB did not reduce the number of matings or the implantation of embryos. However, fewer live pups were born to females in Group B and they weighed 28% less at birth than controls; Group C produced no pups. Five days after birth, 82% of the pups were alive in the controls (A), but only 17% in B. The females were then killed and the amount of PCB in muscle was determined.

If we assume that otters are similarly sensitive to PCBs, then we can compare the tissue PCB levels of otters with those of the mink that suffered depressed reproduction under experimental conditions (Fig. 4.3). The majority of otters from southern Sweden, and half of those from northern Sweden, had tissue concentrations of PCBs above those in experimental mink suffering reproductive impairment; no otters from Norway were at risk. The conclusions seem very compelling.

PCB concentrations in Canadian otter from North America are generally lower than those of Norway (Table 4.2) so that, from the small amount of data available, PCBs there would appear not to be a problem. Nevertheless, PCBs are almost ubiquitous in fish in the United States

(Schmitt, Ludke and Walsh, 1981) and they have been implicated in the decline of otters in Oregon (Henny *et al*., 1981). No data on PCB residues in British otters have been published and Chanin and Jefferies (1978) concluded that PCBs could not have been implicated in the decline, because there was no sudden increase in their manufacture in the mid 1950s. Nevertheless, it may have taken some time for PCBs to reach critical concentrations in otters and a failure in reproduction may have occurred at approximately the same time over a wide area. The limited amount of data on PCBs in British fish (Holden, 1973; Hider, Mason and Bakaj, 1982; Rickard and Dulley, 1983) suggest that levels would not be sufficiently high to interfere with reproduction in otters. However, piscivorous herons (*Ardea cinerea*) contain high PCB concentrations (Prestt, Jefferies and Moore, 1970) and all sparrowhawks analysed by Newton and Haas (1984) contained PCBs, with no decline in levels during the period of study. Clearly the role of PCBs in the decline of the British otter requires further study and, of 14 animals analysed during 1984 (Table 4.2), five contained concentrations of PCB potentially high enough to impair reproduction.

The third major group of toxic pollutants are the heavy metals, about which most concern has been expressed over mercury, cadmium and lead. Mercury has a wide range of uses, especially in the chloralkali, electrical and paint industries and it is used widely as a fungicide, both in industry and agriculture. Most cadmium is used in the electro-plating and plastics industries and in pigments. Over half of the lead is used by the motor industry, in batteries and as an anti-knock agent in petrol. The toxic effects of mercury and lead are mainly directed at the nervous system, while lead also damages the blood and kidneys. Elevated cadmium levels cause kidney damage.

O'Connor and Nielsen (1981) have experimentally investigated mercury poisoning in Canadian otters. They live-trapped otters and fed them mercury in the diet at concentrations of 2, 4 and 8 p.p.m. Hg. All showed clinical signs of mercury poisoning and either died or were killed when symptoms were advanced. The controls showed no symptoms. The concentration of mercury in the liver averaged 33 p.p.m., and a liver concentration down to 25 p.p.m. was associated with marked clinical symptoms. O'Connor and Nielsen concluded that a regular intake of food containing 2 p.p.m. mercury would prove lethal to otters in the wild. Deaths have been recorded in mink receiving a dietary intake of 1·8 p.p.m. Hg (Wobeser, Nielsen and Schiefer, 1976).

Levels of mercury of 2 p.p.m. are on the high side for fish tissues,

Table 4.4. *Concentrations (p.p.m.) of heavy metals in otter tissues*

	No. samples	Tissue	Mercury Mean	Mercury Range	Cadmium Mean	Cadmium Range	Lead Mean	Lead Range	References
Lutra lutra									
Great Britain	24	hair, dw	21·3	1·3–159·4	1·2	0–15·9	14·1	0–88·5	1
Sweden	8	liver, fw	16·5	4·1–30·7					2
	42	muscle, fw	1·8	0·45–9·6					3
Norway	15	muscle, fw	0·57	0·23–1·0					3
Lutra canadensis									
North east U.S.A.	28	liver, fw	1·63	0·34–5·12					4
Georgia, U.S.A.	9	hair, dw	30·4	9·3–67·9					5
Virginia, U.S.A.	357	liver, dw			0·12	0–1·58	2·14	0–55·89	6
	357	kidney, dw			0·55	0–14·9	1·76	0–9·75	6
	276	bone, dw					2·51	0–35·16	6
Wisconsin, U.S.A.	1	hair, dw			0·98		9·4		7
	49	hair, fw	6·47						8
	49	liver, fw	3·34						8
	49	kidney, fw	8·47						8
	49	muscle, fw	1·44						8
	49	brain, fw	0·74						8
Ontario, Canada	4	muscle, fw	0·89						9
		liver, fw	2·97						9
		kidney, fw	2·23						9
Manitoba, Canada	38	liver, fw	2·51	0·52–8·88					10
		kidney, fw	1·69	0·01–6·48					10
		brain, fw	0·52	0·01–3·13					10

Note: dw = dry weight; fw = fresh weight. References: 1, authors' unpublished data; 2, Erlinge, 1972*a*; 3, Olsson *et al.* (1981); 4, O'Connor and Nielsen (1981); 5, Cumbie (1975); 6, Anderson-Bledsoe and Scanlon (1983); 7, Smith and Rongstad (1981); 8, Sheffy and St Amant (1982); 9, Wren *et al.* (1980); 10, Kucera (1983).

but they do occur, with concentrations as high as 10·5 p.p.m. reported in the literature (Förstner and Wittman, 1981; Jernelov and Martin, 1980). Population declines of otters in some areas could be caused by mercury poisoning. It is likely that sublethal effects on otters will occur with diets considerably less than 2 p.p.m. Hg and it is a pity that the experiments of O'Connor and Nielsen reflected only the highest concentrations of mercury recorded in fish populations.

Table 4.4 shows the results of analyses of otter tissues for heavy metals. Mercury concentrations in the liver of Swedish otters ranged up to 30·7 p.p.m. (Erlinge, 1972a), almost as high as the concentration known to be lethal to Canadian otters (33 p.p.m.); there must surely have been sublethal toxicity in this population. In human beings, it is generally considered that minor neurological disorders can be detected in some individuals with hair concentrations of 50 p.p.m. Hg, while the threshold value for pregnant women may be as low as 15 to 20 p.p.m. Hg (see review by Piotrowski and Inskip, 1981). A wild mink, known to have died from mercury poisoning, had a concentration of mercury in its fur of 34·9 p.p.m. (Wobeser, 1976). The mercury levels recorded in otter hair from Georgia are within this range, while 6 out of 24 British otters had levels greater than 20 p.p.m. (Table 4.4, authors' unpublished data). The analysis of hair is a potentially valuable technique for assessing pollution loads, because many elements are concentrated in it (Folio et al., 1982) and hair can be collected from live animals. However, surprisingly little use has been made of hair.

To summarize this section on direct toxicity, the decline in the British otter population coincided with the introduction of cyclodiene pesticides into agriculture, the decline in the Swedish otter population has been linked to high tissue concentrations of PCBs, while mercury levels in some otter populations may be high enough to cause adverse effects. It must be remembered, however, that wild otters may carry a suite of pollutants at concentrations lower than those required to produce overt effects. We do not know how these may be operating, either singly or in combination, to reduce the fitness of otter populations.

Indirect methods of assessing pollution in otters
European otters only rarely become available for toxic chemical analysis. The analysis of fish may help us to assess the potential threats to an existing otter population, or it may provide clues to the disappearance of otters from a river. Furthermore, it is a necessary first step in evaluating the suitability of an area for re-introduction of otters (p. 153).

To assess the potential significance to otters of toxic residues in fish we need to make a number of assumptions. In the discussion below, we have assumed that the average otter consumes 1 kg of fish each day, this value being at the lower end of the range (1–2 kg) of food consumption reported from captive otters (Erlinge, 1968a; Duplaix-Hall, 1975; Wayre, 1979a). We have to assume that otters eat fish in proportion to those in the sample and that the sample is representative of the general contamination of fish at the site. In assessing the possible effects on otters, we must assume that they respond to specific pollutants in a way similar to that of humans or experimental animals, on which data are available.

During 1980 and 1981, more than 500 fishes from 95 sites in Great Britain were analysed for chlorinated hydrocarbons, and analyses for metals were carried out on muscle tissue of over 250 fish from 67 localities (Hider et al., 1982; Mason, Macdonald and Aspden, 1982). The distribution of sites, of necessity, did not represent the current distribution of otters, but rather the distribution of agencies prepared to catch fish. They do allow for the assessment of potential risk to otters. Chlorinated hydrocarbon contamination in fish appears now to be patchy, but there are some areas of concern. The levels were compared with the criteria for the protection of wildlife in the U.S.A. (National Academy of Sciences, 1973), as no such guidelines exist in Britain. Of 37 fishes (6·9%) containing detectable DDE, only one had a concentration above the maximum allowable level of 1 p.p.m. Nineteen fishes (3·5%) contained dieldrin, with 9 (1· 7%) having levels above the guideline of 0·1 p.p.m. It would appear then that organochlorine pesticides are no longer generally distributed in the British freshwater ecosystem in amounts likely to adversely affect otter populations. Nevertheless, there are still local hotspots and, as these compounds are not banned by law, the situation needs continued monitoring. It is also significant that 15·6% of eels, a favoured food of otters, were contaminated with chlorinated hydrocarbons, compared with only 7·4% of other fish. DDE was recovered from all of the eels collected from the River Lyn, in north Devon, whereas none of the trout, collected from the same site on the same day, had measurable residues. Clearly, eels are bioconcentrating pesticides.

Mercury, cadmium and lead (Fig. 4.4) were much more widely distributed than chlorohydrocarbons, but they occur naturally in the environment at low levels, making the interpretation of data more difficult. A species which has always been exposed to metals may have evolved mechanisms to minimize their toxic effect. For want of data, we

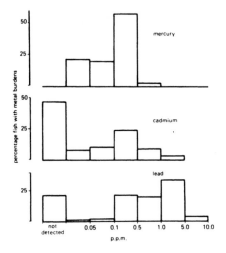

Fig. 4.4. Heavy metal burdens (p.p.m. in muscle tissue, fresh weight) in 259 fish collected from 67 sites in Great Britain.

have assumed that such mechanisms do not exist. It is worth noting here that seabirds off the coasts of north-west Scotland are exposed to naturally elevated cadmium and show high concentrations in their kidneys. The seabird colonies are healthy and expanding and it has been assumed that they must have evolved adaptations to cope with cadmium. However, nephrotoxicity (kidney damage) has recently been detected in these birds (Nicholson *et al.*, 1983); although they were outwardly healthy, they may be disadvantaged during adverse conditions.

The allowable daily intake of mercury for humans, based on F.A.O./W.H.O. recommendations is 0·03 mg, while the threshold level for toxicity would be achieved with a regular daily dietary intake of 0·2 to 0·5 mg (Piotrowski and Inskip, 1981). Using the assumptions made above, the allowable intake for otters would be exceeded at 93% of the sites investigated, whereas the threshold level would be exceeded at 15% of sites.

The allowable daily intake of cadmium for humans is 0·07 mg and physiological damage will occur with a daily intake of 0· 25 mg (Friberg *et al.*, 1974). The tolerable intake for otters would be exceeded at 37% of sites and the critical intake at 15% of sites.

On the basis of information from Waldron (1980), the warning level for physiological damage would be exceeded on a daily intake of lead of 0·35 mg, while an intake of 0·8 mg would be considered dangerous. For otters, these levels would be exceeded at 64% and 40% of sites, respectively.

These assessments appear depressing, but it must be stressed that

Table 4.5. *Concentrations of chlorinated hydrocarbons (geometric means, mg/kg fat) in spraints from the R. Mole, Devon*

Month	No. Spraints	Lindane	Dieldrin	DDE
October 1983	6	12·2	N.D.	8·8
December 1983	10	40·5	37·4	46·8
January 1984	9	27·1	N.D.	23·9
March 1984	21	16·9	21·9	N.D.

Note: N.D., not detected.

many of the sites were outwith the current range of otters, while our assumptions could be open to dispute. We are not suggesting that otters ingesting these concentrations of metals in their diet would show overt signs of poisoning. They *might*, however, suffer sublethal damage, as described above for seabirds, which could be critical to survival under adverse conditions. Several such pollutants might be present in one individual (19% of our sites had at least two metals above warning levels for health) and their combined effects are unknown. Rather low intakes of lead are known to reduce the discriminatory ability and the development of locomotory and exploratory skills in laboratory rats (Winneke, Brockhaus and Baltissen, 1977; Crofton *et al.*, 1980) and such skills are obviously of crucial importance to an otter. To demonstrate such effects in otters in laboratory or field would be extraordinarily difficult.

In the U.S.A., guano samples have recently been collected from caves to estimate the pesticide burden of bats (Clark, 1981; Clark, Laval and Tuttle, 1982). Elevated levels in guano may alert the investigator to possible pesticide problems in the bats themselves. A similar technique may be applicable to otter spraints. A recent (spring 1983) leakage of dieldrin into the River Mole in North Devon resulted in the deaths of many fish (Elizabeth Lenton, pers. comm.) and we were concerned about the potential effects on the otter population. Samples of spraints have been collected and analysed (Table 4.5). The anal gland, its secretions rich in fatty acids, may concentrate and excrete pesticides. The analysis of spraints to detect pesticide burdens in otters clearly has some potential and our recent analysis of spraints from other rivers in England and Wales (authors' unpublished data) has revealed the fairly widespread presence of chlorinated hydrocarbons.

Spraints can also be analysed for metals, though interpretation is much

Table 4.6. *Concentration of metals in otter spraints*[a]

Region	Number of samples	Mercury	Cadmium	Lead
South-West England	165	$0·60 \pm 0·05$	$2·87 \pm 0·12$	$26·14 \pm 0·57$
Wales	68	$0·41 \pm 0·04$	$3·03 \pm 0·14$	$14·62 \pm 0·59$
North-east Scotland	201	$0·88 \pm 0·07$	$1·68 \pm 0·05$	$18·67 \pm 1·45$
South-west Scotland	113	$0·47 \pm 0·04$	$1·87 \pm 0·10$	$17·43 \pm 1·06$
West coast of Scotland	34	$0·31 \pm 0·05$	$4·04 \pm 0·47$	$18·33 \pm 1·09$
North-east Greece	30	$0·29 \pm 0·02$	$2·66 \pm 0·24$	$18·89 \pm 1·72$

[a] (p.p.m. dry weight, means ± standard error).

more difficult than with pesticides, because the metals may be associated with those components of diet that are passed, unassimilated, through the gut. Lead, for example, is concentrated in bone, while fish-scales may concentrate mercury.

Table 4.6 summarizes an analysis for metals of spraints from several parts of Great Britain, with a small sample from Greece. The concentrations are likely to represent background levels, though there are no data with which to compare them. The spraints from the coastal population of otters (Loch Sunart, Scotland) had higher levels of cadmium than river populations; seabirds from western Scotland also contain elevated cadmium concentrations (p. 103). A small sample of mink scats from south-west England and Wales had lower concentrations of mercury and cadmium than the corresponding otter spraints, probably reflecting the minks' lower dependency on fish. The lead level of mink scats was, however, higher than in otter spraints, probably because of their higher content of mammal and bird bones.

Faecal cadmium concentrations have been used to assess dietary intake in man (Kjellström, Borg and Lind, 1978), because very little cadmium is absorbed across the gut. The critical, long-term daily intake of cadmium for man is 0·25 mg. Assuming a similar intake for otters, this would represent a faecal concentration of about 10 p.p.m. Three spraints collected, after a long search, from a remnant, or transient otter population in two Cornish rivers, known to be suffering from long-term pollution from mines, had cadmium concentrations of 5·9, 8·6 and 10·3 p.p.m. Where declines in otters are occurring, or where pollution is expected, it may be well worth analysing spraints for metals and comparing them with the values given in Table 4.6.

Habitat destruction

In almost any account of a rare or threatened species, habitat destruction will be cited as a major cause of decline. The term is used so often that it risks losing impact and frequently there is a lack of understanding of those specific elements within the habitat that are vital to the survival of the species.

Otters in different countries or regions may depend upon differing features of habitat. The animals need shelter in which to sleep and breed and their choice of resting sites appears catholic. Holes in the river bank, cavities amongst tree roots, piles of rock, wood or debris may all be used. Otters will sometimes enlarge rabbit burrows or excavate tunnels in rooted tree stumps. Field drains and broken masonry offer man-made places of shelter, while dense riparian vegetation can also prove ideal.

In many parts of Britain otter holts can be found within the root systems of mature bankside trees. Initial studies on the River Teme catchment in western England (Macdonald *et al.*, 1978) suggested that otter distribution related to the availability of possible resting sites, to the density of bankside vegetation and to the density of mature ash (*Fraxinus excelsior*) and sycamore (*Acer pseudoplatanus*) trees. This river system apparently contained a low, remnant population of otters (p. 62).

This work was later extended to cover a wider variety of river types in Wales and the border counties (Macdonald and Mason, 1983*c*). Fifty 5 km stretches of rivers were walked and the number of sites at which evidence of otters was found, together with the number of spraints, were noted. Data were also collected on 15 parameters of habitat considered of possible importance to otters. These factors included the total number and type of bankside trees, numbers of potential resting sites, the presence of mink, information on water quality and fish stocks and an index of human disturbance. The data were subjected to multivariate analysis and it was found that the distributions of positive sites and of otter signs were correlated with the presence of potential resting sites and with, once again, mature ash and sycamore trees. Many of the holts located by Green *et al.* (1984) while radio-tracking otters in Scotland were situated in the roots of ash, sycamore and elm.

Fig. 4.5. The roots of (a) ash and (b) sycamore tend to grow above the water table, horizontally, allowing erosion by water to form large cavities. By contrast, the roots of alder (c) and willow seek water, forming a dense, fibrous mat, with erosion much reduced; otter holts are infrequent beneath these species.

The reason for the importance of certain tree species in offering holt sites for otters lies in the structure of their root systems (Fig. 4.5). Bankside ash and sycamore develop open root systems that, to avoid waterlogging, tend to grow horizontally. Oak (*Quercus* spp.) roots have a similar structure and in a field survey of the River Teifi in west Wales 13 holts were located amongst oak roots (Brooker, 1983). In western Britain, the majority of rivers flow quite swiftly and are subject to spates. In conditions of high water, soil is eroded from between the roots

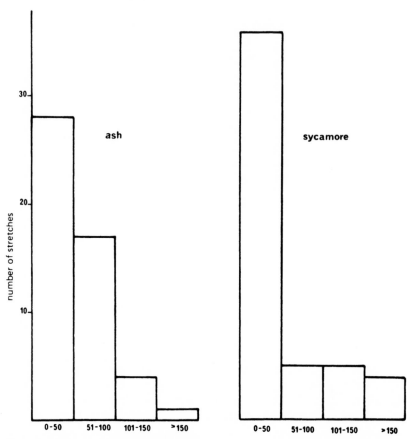

Fig. 4.6. Frequency distribution of numbers of mature ash and sycamore trees on 50 5 km stretches of river. (From Macdonald and Mason, 1983c.)

leading, eventually, to the formation of cavities within the bank. As the tree ages, it tends to tilt towards the water and extended cavities are formed. Where several such trees grow closely, underground tunnels and chambers may ramify extensively. In such places the otter finds security.

Such trees are now much less common than in the past and in many places are at a premium (Fig. 4.6). They have been removed by water resource managers. It is the brief of the regional water authorities in Britain to improve land drainage and to reduce possible risks of flooding. A large tree toppling into a river forms a barrier leading to large accumulations of floating debris and additional risks of flooding. Large trunks carried downstream by spates can damage bridges and private property. So, mature trees, that seem likely to fall, are systematically removed. Bankside vegetation is also cleared to allow access to heavy

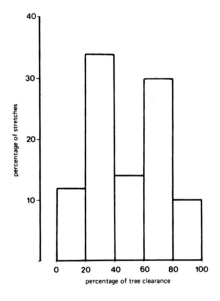

Fig. 4.7. Tree clearance along 50 5 km
stretches of river. Maximum mature tree
density is about 1 tree per 10 m of bank
and the histogram represents frequency
classes of tree densities below the maximum. The histogram is bimodal because
surveys were carried out on both tributaries
and those rivers classified by water
authorities as 'main river', which receive
more intensive management.

machinery where channels are to be modified for improved land
drainage.

The root systems of several tree species are sometimes suitable for
otters and holts have been found in, for example, holly (*Ilex aquifolium*),
beech (*Fagus sylvatica*), and lime (*Tilia cordata*). Ash, oak and sycamore
do, however, seem to be of prime importance.

On the majority of British rivers, and especially in the lowlands, the
predominant tree species are willow (*Salix* spp.) and alder (*Alnus
glutinosa*). These trees have dense, fibrous, water-seeking roots which
are often impenetrable to otters, although they are sometimes used.
Harper (1981), for example, found holts used for breeding under alders
in Scotland and Erlinge (1967*b*) mentions them as den sites in Sweden.
Willow and alder roots are ideal for the prevention of bank erosion and
their canopies provide valuable general cover along waterways. Ideally,
a mixture of tree species along river banks will increase both bank
stability and ecological diversity.

In our survey of rivers, the number of mature trees recorded for each
5 km stretch varied between 1029 and 3. The higher number was found
on the River Onny in Shropshire, with a diversity that represents a
woodland remnant association (Mason, Macdonald and Hussey, 1984).
Only three trees were recorded on a stretch of the River Severn that had
been heavily managed for flood prevention and that now offers no cover
for otters (Fig. 4.7).

Fig. 4.8. This stream (the Ellington Brook), near Huntington, Cambridgeshire, has been dredged and both banks cleared of vegetation. It retains no wildlife interest.

As a whole, in Britain, the rivers in the lowlands of central, eastern and southern England have been most heavily managed (Fig. 4.8). This may well be a contributory factor in the present rarity or absence of otters from these parts of the country (see p. 60). The dramatic loss of riparian trees can be exemplified by the situation in Essex, eastern England. A comparison of old maps with aerial photographs was used to demonstrate the rate of loss in the Rivers Colne, Blackwater and Stour (Fig. 4.9). The River Stour was navigable during the last century as far as Sudbury, so stretches above and below this town are treated separately. It is apparent that extensive tree removal had occurred prior to 1879, especially on the navigable Stour, using a yardstick of 200 trees/km (both banks) as was found on the little-managed Onny (see also lines from John Clare at the head of this chapter). The rate of tree clearance along the River Colne, which still had reasonable cover in 1879, was 10% per decade, over the period studied. The average tree density that appears acceptable in Essex to watercourse managers with current attitudes is 25 to 35 trees/km.

Today the otter is extinct in Essex but the effects of tree removal on its distribution can be seen on the River Severn. The Severn rises in mid-Wales and flows through western England to the Bristol Channel. Otters are resident only in the upper reaches of the river and from regular monitoring it is possible to identify the present downstream limit to their regular distribution. Within 25 km both upstream and downstream of this point, five random 1 km stretches were walked and all trees recorded. As can be seen from Fig. 4.10, far greater numbers of mature

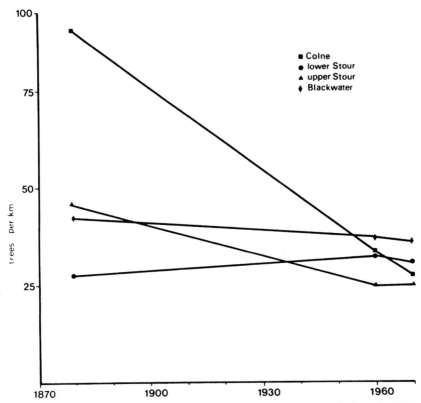

Fig. 4.9. The loss of trees along Essex rivers. All riparian trees marked on the 1879 Ordnance Survey maps (scale 6″ to 1 mile) from source to tidal limit were counted, as were trees shown on aerial photographs (scale 1 : 10 000 at 6000 feet) taken in 1960 and 1970. A 'natural' stretch of river would support about 200 trees per km, emphasizing that extensive modification of the riverside vegetation on these East Anglian rivers had taken place by the last century, if not before.

trees were found on those stretches inhabited by otters. On the downstream stretches, overall numbers of trees are lower, and the proportion of young trees is greater. These are mostly *Salix* saplings. The diversity of species is also largely lost on the heavily managed reaches (Fig. 4.11).

A lack of trees is not, however, always detrimental to otters so long as other forms of shelter are readily available. On the coasts of Scotland, Norway and Portugal, for example, holts (Fig. 4.12) are found amongst rock falls (Kruuk and Hewson, 1978; Macdonald and Mason, 1980; Simões, 1977–1982). Similar sites can be found on Portuguese and Spanish rivers (Macdonald and Mason, 1982b; Elliot, 1983). Dense

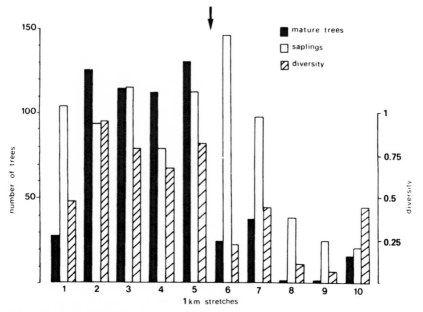

Fig. 4.10. Counts of bankside trees were made on 10 1 km stretches of the River Severn. Data to the right of the arrow derive from stretches downstream of the regular range of the local otter population.

scrub also provides cover (Fig. 4.13). Geoff Liles knows of otters breeding in thickets of bramble (*Rubus* agg.) and blackthorn (*Prunus spinosa*) in west Wales. Dry 'islands' of bramble within reed beds are favoured on the Norfolk coastal marshes and similar sites are used by otters in the marshes of Iraq (Thesiger, 1964). Impenetrable *Salix* scrub offers shelter on Greek rivers, whereas in southern Italy the *Salix* is regularly cut or is destroyed by gravel workings. While radio–tracking otters Green *et al.* (1984) found many resting sites situated above ground. Some of these were simply depressions in vegetation no more than 30 cm high, but most were associated with *Salix* scrub, *Rhododendron* or *Polygonum cuspidatum*. It is important that the riparian vegetation be dense and that an otter, once disturbed, can slip away unseen to find another refuge.

In the absence of trees or scrub, marshland with *Phragmites* or tussock sedge provides ideal cover for otters. Lyn Jenkins surveyed the catchment of the Western Cleddau in west Wales, which still contains wild marshes in its upper reaches. She recorded all signs of otters and noted both bankside and hinterland habitats (Jenkins, 1982). The downstream

Fig. 4.11. Vegetation along the River Severn: (*a*) dense bankside trees in a stretch frequented by a good otter population; (*b*) a heavily managed stretch, supporting only a few scrubby willows, in which otters occur only at irregular intervals.

sections of the river have been heavily managed. Trees have been cleared and wetlands drained to improve the farmland. Jenkins found that otter activity was significantly higher on tributaries that were still in a near-natural state than on those which had been managed. There was a positive association between spraint density and good vegetation cover, especially where woodland or *Rhododendron* thickets were present.

Fig. 4.12. Typical rocky habitat (a) on the west coast of Scotland containing a well used otter holt (b).

Fig. 4.13. Otters will lie up in dense bankside scrub such as this, by the River Severn, from which an otter was flushed.

Where pasture had been improved and banks left bare, far fewer otter signs were found. She suggests that the distribution of otters within the catchment has been affected by losses of wetlands adjacent to the river and of woodland.

Jenkins and Burrows (1980), and Bas, Jenkins and Rothery (1984) working in north-east Scotland, also found that spuanting activity by otters correlated with the presence of riparian tree rows and with bankside woodland. In Britain, since 1949, 30–50% of the ancient lowland woods have been destroyed (Nature Conservancy Council, 1984) and many have been converted to commercial conifer plantations. Although conifers may ultimately affect the food supplies of otters (Smith, 1980), young plantations do offer shelter for the animals away from the river. Woodland with a vigorous undercover is also valuable.

Neither woodlands nor tree rows are characteristic of the Somerset Levels in south-west England. This flat wetland, intersected by small water courses and prone to winter flooding, is of national importance for its wildfowl and flora. Not so long ago the Levels were considered a stronghold for the otter but numbers are now dwindling. Following a field survey of the region, Jarman (1979) suggested that areas of greatest agricultural improvement were lacking otters. Elizabeth Lenton has monitored the region since the early 1970s and comparisons of her results give some measure of the decline. In 1977, for example, she visited 195 sites and found signs of otters at 53. In 1983, signs were found at only 19 of those same sites.

In the Somerset Levels (Fig. 4.14), prime otter habitat is continuously being lost. Traditionally, peat has been extracted from the valleys of the inland rivers and small peat workings can provide ideal otter habitat. *Phalaris*, *Carex*, *Typha* and *Salix* scrub offer cover, and holt sites may be found in old peat banks. Between 1976 and 1983 the production of peat from the Levels doubled and the large-scale commercial workings quickly destroy extensive areas of vegetation cover. Most of the peat is used in horticulture (Lenton, 1985).

Huge tracts of the Somerset Levels have also been drained for the conversion of permanent pasture to arable land. In Britain, land drainage schemes for agricultural improvement are subsidized by the Government and public expenditure on drainage in 1980/1981 amounted to £152 m of which £69.2 m was classed as capital (improvement) expenditure (Royal Society for the Protection of Birds, 1983). Recently, continuing drainage of the Levels has resulted in fierce argument between farming and conservation lobbies. West Sedgemoor, for example, is one of the

Fig. 4.14. In the Somerset Levels, a wetland of international importance, peat extraction and land drainage are destroying otter habitat.

richest areas for wildlife, amounting to 1000 ha. The Ministry of Agriculture, Fisheries and Food grant-aided 18 capital land drainage schemes on West Sedgemoor, despite opposition from the Government's own advisory body, the Nature Conservancy Council. The rapid destruction of the Somerset Levels threatens many forms of wildlife and the future of the otter in this region now looks very bleak.

With some understanding of the specific features of wetlands that are utilized by otters, it is possible, even during brief field surveys, to estimate the quality of habitat at different sites. As outlined in Chapter 3, short surveys have now been carried out in several southern European countries. Some comparison can be made between the status of the otter and the availability of suitable habitat in those countries (see Table 4.7).

While it is possible to relate otter status to our measures of habitat quality, it is far more difficult to investigate habitat loss experimentally. In Britain, at a local level, pioneer tree clearance by a water authority may be highly destructive, but resident otters will continue to mark the area if it is a part of their normal range. Sufficient shelter may still be available on neighbouring stretches and, since there is no direct relationship between numbers of spraints found and actual numbers of otters, interpretations are difficult. Any effect on the population could only be seen in the long-term and, even then, a decline could be due to many other causes.

In some places, interpretation may be easier because extensive regions of otter habitat have been virtually eliminated! In the coastal lowlands

Table 4.7. *The percentage of sites offering good, moderate and poor cover for otters in Portugal, Greece, Spain and Italy*

Country	Good	Moder-ate	Poor	No. sites surveyed	% sites positive
Portugal	41	37	22	90	70
Greece	33	42	25	200	62
Spain	36	35	29	176	40
Italy	18	34	48	188	8

of Morocco, for example, few rivers now flow between Casablanca and Agadir or inland to Marrakesh. Since 1955, many dams have been built in the uplands (Abouzaid and Hajji, 1982), where waters are retained, leaving dry river beds in the lowlands. Water for irrigation is frequently supplied in raised concrete channels. The otter in Morocco is now mostly to be found in rivers flowing through low hill country with dense bankside vegetation.

Deforestation too affects water flow and is an international problem. Throughout southern Europe, for example, torrents may pound down hillsides in the winter but in summer those same river beds are dry. Even where waters flow all year, erosion resulting from deforestation or overgrazing leads to high silt levels in waters. In places such as southern Italy, extraction of gravel from river beds produces the same effect. Where levels of suspended sediments are high, fish and crustacean stocks are severely depleted.

Fish stocks can also be affected by a lack of bankside trees. Shading helps to maintain water temperatures and oxygen levels that are suitable for salmonids (Meehan, Swanson and Sedell, 1977; Karr and Schlosser, 1978). Underwater roots provide hiding and feeding places for fish. In a situation where instream cover of this sort had been removed during management of an English river, Swales (1982) found that the mean standing crop of the fish population fell by 76%. We have no information, however, about the effects of such reductions on resident otter populations.

It can be seen that habitat destruction can affect otter populations in a variety of ways. However, while it is possible to identify some of the features that are significant, we still do not understand the minimum habitat requirements of *Lutra lutra*. Only with this information will it be possible to predict the impact of varying levels of habitat alteration.

Within its range, how many resting sites does an otter use? Green *et al.* (1984) found that three otters used 47 different sites that were not specific to individual animals; resting places were usually changed daily. A male used 29 sites, but one female used five and another 14. The authors differentiate between holts, a term they use for underground dens, and couches, which comprise any sites above ground, including stick piles. They found that the average distance between resting sites was 1532 m of waterway, but sites were clustered at centres of otter activity. Destruction of habitat within activity centres may prove more damaging than elsewhere. Activity centres themselves may be chosen in relation to food resources. Green *et al.* suggest that, since they found that female otters select more secure resting sites than males, any limit to such sites might have more effect on females and on breeding. The only holt that they describe as being used for breeding was in rock 40 m from water and free from any threat of flooding, and Harper (1981) also found breeding holts to be situated above flood levels. Within Green *et al.*'s study area, there is no evidence that resting sites are in short supply.

On the west coast of Scotland, Jane Twelves (pers. comm.), also by radio-tracking, found that individual otters used only one holt site for much of the time, with subsidiary holts being used only occasionally. In her study area otters are very common.

In Ireland, where otters are also common, their signs can be found on waterways apparently offering no cover. The animals may use these streams only for feeding or it may be that young otters are relegated to sub-optimal habitat. However, the high level of recruitment in Ireland must reflect the presence of suitable habitat along most rivers. Radio-tracking of otters in sub-optimal habitat and in low, fragmented populations is now required.

Finally, it must be stressed that, although habitat destruction can affect otters in a number of ways, there are now many areas where habitat appears to be ideal but where otters no longer occur. Regions like this can be found throughout Europe but perhaps one of the most dramatic examples is the Camargue in southern France. Considered as one of the finest wetlands and one of the last true wildernesses of Europe, the otter has not been recorded in the Camargue since the 1950s (Luc Hoffmann, pers. comm.). The reasons for its demise are not known but certainly a lack of habitat was not a contributory factor.

Competition

In Britain the only exotic carnivore to have become established in the wild is the North American mink, *Mustela vison*. The species is now also common in much of northern Europe. In the U.K., animals originally escaped from fur farms or were released by failed entrepreneurs. Breeding in the wild was first recorded in Devon, south-west England, but at much the same time animals were reported from Suffolk in the east and from northern Scotland (Lever, 1977). Today, mink are present throughout much of England, Scotland and Wales and also occur in Ireland. The spread of mink in Britain caused an outcry from landowners and from some conservationists, leading to attempts by Government agencies to eradicate the species, an attempt that failed. The mink was, and still is, widely condemned as a voracious killer and there have been problems with losses of ground nesting sea-birds on offshore islands (Bourne, 1978). When it became clear that the otter had declined over much of England, the mink was automatically cited as a culprit (e.g. Lever, 1978).

It now must be accepted that *Mustela vison* is in Britain to stay and work has been carried out on the ecological relationships between it and the otter. Since mink, like otters, are found in aquatic habitats, comparisons have been made of their feeding biologies.

Erlinge (1969, 1972*b*) was the first to compare the diets of otters and mink. On a Swedish river, otters ate over 90% fish, whereas fish comprised only 60% of the diet of mink, which ate many more mammals and birds. On a lake, otters were again fish specialists, while mink took many mammals, with crayfish being a shared resource. Overall, there appeared to be a 50% overlap in summer diet and a 70% overlap in winter, when some foods became unavailable due to ice cover. Competition might occur for limited resources in winter and Erlinge considered that mink might be excluded from the preferred habitat of otter when otter numbers are high.

Comparisons of the diets of the two species have recently been carried out in the milder climate of south-west England. Wise *et al.* (1981) examined in detail the diets of otter and mink on a eutrophic lake and an oligotrophic moorland river system and their overall comparisons are summarized in Fig. 4.15. As in Sweden, fish formed the main diet of otter in both types of habitat, while the diet of mink was much more catholic. Overlap between the diets was greatest in autumn and winter, when mink took more fish and the authors conclude that competition

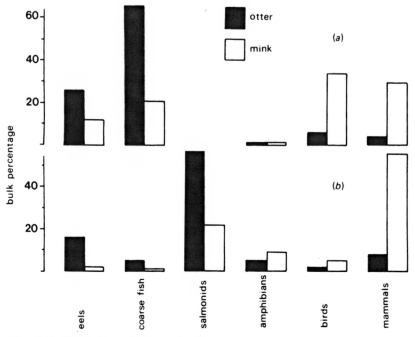

Fig. 4.15. The bulk percentages of main prey items taken by otters and mink on (a) a eutrophic lake and (b) an oligotrophic river in Devon, south-west England. (Adapted from Wise *et al.*, 1981.)

for food is unlikely. Chanin (1981), in a study of the same lake, suggested that, although dietary overlap was such that competition could not be excluded, the presence of the otters may have kept the mink population low. Birks and Linn (1982), who also worked in this region, found that, in oligotrophic waters where aquatic prey densities were low, foraging by mink was more terrestrial. Jenkins and Harper (1980), analysing scats from both otter and mink in north-east Scotland, also confirmed the differences in diet between the two species. It seems clear from all these studies that the otter is a fish specialist, while the mink is an opportunist and a generalist.

In North America, where *Mustela vison* is endemic, it co-exists with *Lutra canadensis*. Melquist, Whitman and Hornocker (1981), after radio-tracking 26 mink and 37 otters in Idaho, concluded that, although there was some dietary overlap involving a few fish species, effects were minimal because food was not limiting. The two carnivores also displayed differing foraging strategies. Food supplies did not become locally depleted, because the otter moved from site to site while foraging. Otters could take fish that were too large for mink to handle and mink took substantially more terrestrial prey items. Both animals occupied the same log jams on rivers, which often occur in otter activity centres, being associated with high prey densities (Melquist and Hornocker, 1983). Within the log jams, mink were able to live in places that were inaccessible to otters. The American biologists concluded that the two species could co-exist because of the partitioning of resources.

In Britain, Jim and Rosemary Green (pers. comm.) have recorded mink using the same holt site as *L. lutra*. Birks and Linn (1982) found that mink dens were commonly located in the root systems or in hollows within waterside trees, especially oak, willow and sycamore. Rabbit burrows were also frequently used by mink and, after enlargement, these are also used on occasion by otters. Mason and Macdonald (1983), working largely in Wales, found that the distribution of mink signs correlated with the density of willows and general bankside scrub, the type of cover also considered important to mink in Sweden by Gerell (1967). In Britain, mink frequently occur in habitat that is sub-optimal for otters and, to date, there is no evidence to suggest that mink are able to evict otters from their holts or prevent otters from using suitable stretches of waterway. Equally, there is no evidence that mink prey on otter cubs. Indeed, Novikov (1956) reported that the otter in Russia 'vigorously hunts mink'.

In Britain, mink only spread extensively after the otter had already declined (Chanin and Jefferies, 1978) and in Sweden too, according to Erlinge (1972b), game records show that the decline of the otter preceded the increase in mink populations. In neither country can mink be cited as a cause for losses of otters and in Britain we have no evidence that mink are preventing recovery of otter numbers.

In this country, mink were reported on the River Teifi and throughout Pembrokeshire in west Wales in the late 1950s and early 1960s. This was thus one of the first areas in which mink became established. Today, mink are still found on these waterways and yet this region is considered to be a stronghold for the otter in Wales (see p. 59). By contrast, in Norfolk, where mink are very rare, the otter is still in decline. In our study of rivers in Wales and the border counties, we could find no relationship between the presence of mink and the presence of otters.

Persecution

In some parts of Europe *L. lutra* may still be subject to natural predation. In Poland, Krzysztof Wlodeck (pers. comm.) reported that otters are sometimes killed by lynx and wolves, and Love (1983) recounts tales of otters being attacked by sea eagles (*Haliaeetus albicilla*) in Scandinavia. It is unlikely, however, that such predation will affect the status of otter populations. Throughout much of western Europe carnivores capable of taking otters have themselves been largely eliminated by man and it is now man himself who is the major 'predator' of the otter.

In the past otters were widely trapped for fur, shot as vermin or hunted for sport. In Britain, otter hunting with hounds was only finally banned in 1981 (Fig. 4.16). Such hunting was not responsible for the initial sharp decline in otter numbers but, once populations were low, the additional pressures of hunting may have been detrimental. Our suggestion (in Macdonald and Mason, 1976) that, in areas where the animals were uncommon, hunts might concentrate on rivers still holding otters was confirmed by the study of hunt records by Chanin and Jefferies (1978).

Today the otter in Europe has been afforded some form of protection throughout much of its range (see Chapter 5) but persecution still continues. At the Third International Otter Symposium held in 1983, several delegates cited poaching as a cause for concern. Wlodeck (1985) considers this to be a prime contributor to the present low levels of otter populations in much of Poland. In Bulgaria, Finland and Portugal, illegal killing was also thought to be a problem. It must be remembered, however, that it is almost impossible to collect accurate figures of numbers of animals killed illegally and there are no studies showing the real effects on distribution or status in Europe.

Since otters were thought to compete with man for fish stocks, they were widely regarded as vermin. In Yugoslavia, for example, high bounties were paid to fishermen until 1966 (Hönigsfeld and Adamič, 1983). The Foyle Fisheries Commission in Ireland operated a bounty system until 1968 and, over a period of almost 11 years, 295 otters were killed in their area (Chapman and Chapman, 1982). Today in Bulgaria, much of the illegal trapping is carried out around fish farms.

Many otters are also accidentally drowned in fish traps of various types. During a visit to Greece in 1983, a fisherman showed us a fyke net (see Fig. 4.17) in which he had caught two otters on the River Strymon. While only relatively small numbers of otter mortalities may be reported to interested authorities, the proportions of deaths by drowning are significant.

Fig. 4.16. *The Otter Hunt* (Laing Art Gallery, Newcastle upon Tyne) (1844) by Sir Edwin Landseer. Ormond (1981) describes the picture as 'the culmination of the chase during which the otter would have been pursued, perhaps for several hours, as it swam up and down the shallows until forced out of its final refuge to be speared, still in the water, by the huntsman. As he thrusts, the hunter twists the spear to ensure the creature is firmly impaled and prevented from writhing off. As if to emphasize its notorious cunning and viciousness, which, together with its depredation of fish stocks, account for the hunters' hatred for the creature, the pierced otter is shown vengefully thrusting against the top of the spear and arching backward to bite the shaft'. Note the two salmon on the bank to confirm the otter's greed and refer back to p. 15 for an assessment of diet. The use of the spear died out at the end of the nineteenth century, but otter hunting itself continued until 1981, when otters received legal protection. Hunters have turned their attention to feral mink, which are similarly 'vicious, vengeful and greedy'.

Fig. 4.17. Fyke nets, used especially for catching eels, are the cause of many otter deaths.

Amy Lightfoot spent eight months studying otters on the island of Hitra off the Norwegian coast. In 1980 and 1981, she received records of 23 dead otters of which 80% had drowned in bow (fyke) nets. Myreberget and Fröiland (1972) had considered hunting to be an important factor limiting otter populations in Norway, but Lightfoot suggests that accidental drownings may have posed greater threats. The nets were often placed in shallow water amongst kelp (*Laminaria*) beds close to the shore, just the type of habitat favoured by otters for foraging (p. 13).

In East Germany, Stubbe (1980) recorded details of 486 dead otters, of which 36% had been caught in fishing nets, while in Poland (Wlodeck, 1980) 31% of 192 otters had also been drowned. In Holland, too, Van Wijngaarden and van de Peppel (1970) reported that, of 979 dead otters, 114 had drowned in bow nets.

In Britain, otters are also killed in fyke nets set for eels. Jefferies, Green and Green (1984) claim that, between 1975 and 1984, more than 80 otters drowned in British eel fyke nets and parlour lobster creels. During the summers of 1975/1976, 23 animals were drowned in nets in small freshwater lochs in the Outer Hebrides (Jane Twelves, pers. comm.). Although otter numbers are high in these Scottish islands, Twelves considered that continuing mortalities at such a level would have been damaging to local populations. In regions where otter numbers are already low, accidental losses could also assume significance. At Leighton Moss, a lake in north-west England, few otters now remain and they are rare in the surrounding district. Recently, two otters, one a lactating female, were drowned in the same fyke net within three weeks. It is likely that this loss represented a high proportion of the annual recruitment to that region. It seems that otters are attracted to the fish already trapped in the nets and such effective 'otter traps' may help to eliminate remnant populations. Norwegian fishermen considered bow nets to be an effective method of catching otters.

Twelves (1983) has also provided data on mortalities in lobster creels. Of 22 otters known to have drowned, the majority were adult females. Twelves suggests that this may reflect unequal sex ratios of otters in the area or the fact that adult males were too large to squeeze through the 12 cm diameter entrance to the creel. Most of the otters that drowned had been foraging in waters less than 4 m deep, although one male became trapped at a depth of 15 m. Twelves considers that, as with fyke nets, the otters are attracted to the live fish that become trapped in the creels. In parallel to the Norwegian situation, she also found that lobster creels that caught otters were often set close to shore and in areas of sea-weed. By contrast, crab creels, which are set on sandy bottoms, catch few otters, probably because otters spend little time foraging over bare sand (see also Watson, 1978).

Traps set by man are seldom selective in the species they catch. Throughout large areas of Europe, the North American musk rat *Ondatra zibethicus* is now well established and, because it damages river banks, there are widespread efforts to control its numbers. Bouchardy (1981) recorded three otters being taken in musk rat traps, while, in Poland, Wlodeck (1980) claimed that otters were not only accidentally trapped but were also hunted due to confusion in the identification of the two species. Live-traps set for coypu (*Myocastor coypus*) have also been known to catch otters in eastern England. The real effect of such mortality on otter status and distribution is not known.

Disturbance

In most general accounts of the European otter, it is stated that the animal is highly sensitive to human disturbance and, at one time, disturbance was considered as a possible reason for declines (e.g. Joint

Fig. 4.18. Small caravan sites, such as this one by the River Severn, occur on many river banks in Wales.

Otter Group, 1977). In our account of the otter on the River Teme (Macdonald *et al.*, 1978), we postulated that a riverside caravan site may have caused sufficient night-time disturbance to deter otters from marking that section of the river.

Figure 4.18 shows a typical small caravan site on the banks of the Severn in mid-Wales. Over three years we have monitored sprainting sites along the river and, while otters mark the bank throughout the year, fewer sprainting sites opposite the caravans are marked in July, August and September. However, the animals continue to mark the banks all year where the same river flows through a town. Chapman and Chapman (1982) regularly monitored sprainting sites on the River Shannon in the centre of Athlone (Eire) and found that, while otters sprainted on stone revetments in the town during two winters, no signs were found between mid-April and October even though the animals were still resident in the area. From these kinds of observations it could be suggested that, although disturbance alone will not deter otters from using a stretch of river, it may affect their marking behaviour. However, where otters are thriving on a river, there seems to be a seasonal pattern of marking behaviour. The number of sites marked with spraints is often lower in the summer even in places away from human disturbance (see p. 33) and this makes it difficult to isolate the disturbance factor as affecting marking behaviour.

In recent years, it has become clear that otters will tolerate a certain level of disturbance so long as the habitat offers sufficient shelter to allow immediate escape. Green and Green (1980), during their survey of Scotland, found signs well within the city boundaries of Glasgow and Aberdeen. In Wales, we know of holts located next to houses. In Ireland, otters frequent loughs that are used heavily for recreation and in central Portugal high levels of disturbance from people washing and bathing or from riverside gypsy camps had no obvious effect on otter distribution. Most disturbance will, of course, occur during the day and, since otters in most places are largely nocturnal, night-time disturbance may be more critical. On the coasts of Scotland, however, the animals are much more active during the day and even here they appear to tolerate human proximity. We have watched an otter floating on its back while playing with a flounder even though there were several fishing boats close by. In these coastal situations, however, the animal can vanish immediately, submerging beneath the sea-weed or finding shelter among rocks.

Results from radio-tracking also suggest that otters are not so sensitive to disturbance as was once supposed. Green et al. (1984) followed an otter while it foraged along a river, despite people walking and fishing and the same otter sometimes spent the day in rosebay willowherb (*Chamaenerion angustifolium*) bordering a village street. Melquist and Hornocker (1983) radio-tracking *L. canadensis* also found that, so long as food and shelter were adequate, otters had a high level of tolerance to human disturbance. Just as in Ireland, the American otters frequented lakes regularly used for recreation and spent time within the city limits of McCall. Once again the determining factor appeared to be the availability of adequate 'escape shelter'.

Unless animals are being intensively persecuted, disturbance alone does not appear to be a factor limiting their populations. However, in much of western Europe today, bankside vegetation does not offer sufficient shelter and, without that security, human disturbance poses a threat. It is the combination of poor habitat, together with disturbance, that may be critical. In many countries, the riparian habitat is poorest in lowlands, since these are the areas most used for agriculture. They are also the regions where most people live and lowland rivers are more likely to be polluted by agricultural and industrial wastes. Elliot (1983), in his survey of Spain, found that otters were more common in upland areas away from human disturbance, but the upland rivers were also less polluted. It is not easy to isolate individual factors as prime causes for otter declines.

Road kills

Like so many wild creatures, otters are killed by road traffic. On some roads there are black-spots where otters are frequently killed at the same place. There was one such site on the River Glaven in Norfolk and Paul Chanin (pers. comm.), working in south-west England, knew of five otters killed at the same place within six years. In such cases it is assumed that the animals were following traditional routes but, where these sites are close to bridges, it is hard to understand why the animals do not pass beneath the roads in the water. It is impossible to estimate real numbers of otters killed on roads since, as with those drowned in fish traps, only a certain proportion are ever reported. Reports of otters killed on roads have come from many countries, be it the Scottish Outer Hebrides, Poland or Czechoslovakia. In West Germany, Heidemann (1981) reported that 26 of 50 dead otters had been road casualties, while in East Germany Stubbe (1980) found that 51 of 486 animals had been killed by traffic. Van Wijngaarden and van de Peppel (1970) reported 55 known traffic victims in Holland between 1940 and the early 1960s. Jenkins (1980) lost at least three otters in this way within one year in his small rural study area in north-east Scotland. It is impossible, however, to predict the effects of these mortalities on the status of local populations, unless perhaps the populations are so low that any accidental death could be critical.

Other factors that may be responsible for losses of otters include severe winters, disease and the non-viability of remnant populations.

When waters freeze, otters are often able to make holes in the ice and so maintain access to their feeding grounds, although Stubbe (1980) reports seven otters being drowned under ice. In Sweden, as Erlinge (1967b) points out, there is ice cover for three to five months of the year, a critical period for the animals. In his study area, the otters left the frozen lakes to congregate along the streams, where open water could still be found. Winter migrations are also reported from Bulgaria, with otters leaving the hills at this time.

Working in north-east Scotland, Jenkins (1980) felt that February was probably a critical month, when a peak of adult mortality occurred. He found that, in severe winters, cubs which had been born during the autumn or earlier in the winter seemed to disappear and only those that had been born before or around mid-summer survived. Jenkins linked his observations to food shortages during periods of hard weather.

Clearly, otters living in a marine habitat will fare rather better. Even on Norwegian fjords covered with ice, there is usually access to the water due to tidal movements along the edge. Availability of fresh water for drinking, and possibly for washing, may be limited (Amy Lightfoot, pers. comm.).

There is little information on disease in otters. Don Jefferies (pers. comm.) has observed blindness in British otters. Stephens (1957) and Harris (1968) both list species of platyhelminthes, nematodes and protozoans that have been recorded from otters. In captivity they appear to be susceptible to problems with kidney stones.

In some regions, where severe declines in otter numbers have occurred, only small and isolated remnant populations remain. As these groups continue to dwindle it is tempting to assume that such low populations are no longer viable and for this reason there is no recovery of numbers. In Chapter 3, we refer to the steady decline that has taken place in the River Teme catchment in western England. The Teme system is rather isolated but, given that otters can travel between watersheds, it is not too distant from other rivers still holding good populations. There is no obvious reason for the continuing decline in the Teme and, while non-viability of fragmented populations could be proposed, it is not proven.

Conclusion

Because of the suddenness of the decline in otter populations over a wide area, pollution would appear to be the major agent in the otter's demise. Opinions differ as to the main pollutant responsible and there is no reason to assume that the same chemical was involved in all areas, or indeed that only one pollutant was involved. Similarly, factors in combination may be responsible for local failures in recovery. During the period of decline there has also been widespread destruction of cover along waterways in the interests of 'river improvement' and for the conversion of land for agricultural use. Large stretches of river are now only of marginal value to otters and this may limit the potential for increases in distribution. Small remnant populations of otters are probably susceptible to other stresses – illegal killings or accidental deaths on roads and in fish traps – which, in better times, they were more able to withstand.

We are still ignorant of many aspects of the biology of otters but we must use what knowledge we have to formulate and put into practice strategies for conserving our remaining populations. This theme will be developed in the following chapter.

5 CONSERVATION

Religione è sollievo ai sofferenti

FRANCESCO MARIA PIAVE, *La Traviata*

Thus far we have described what is known of the biology of the European otter and have examined those factors that have been responsible for its decline. It is clear that, although there has been considerable recent interest in the species, we are still ignorant of many aspects of otter ecology. Most of the detailed research has been carried out on a few individuals in a few localities, but they have suggested that there may be behavioural and ecological variation. Those otter populations in most need of conservation – the small, fragmented and isolated ones – are even less well known; they offer few rewards to the research ecologist in return for long hours of study.

The prudent ecologist might argue that, because we have limited knowledge about otters, practical conservation measures are premature. More research is needed. However, the otter has declined catastrophically in many areas and, for the conservationist, it is essential to halt this decline and to attempt to reverse it. As Bradshaw (1982) has argued, 'In the real world problems often have to be solved in a hurry and with inadequate evidence. We must be prepared to undertake problems even where we realize our answers will be less than perfect'. Our conservation advice will have limitations and we must be prepared to modify it in the light of new knowledge, but we cannot afford to wait until our knowledge is such that we can write a conservation strategy on a tablet of stone.

We can, however, recognize with some confidence certain environmental factors that will affect the success of otters. They require a food supply, especially of fish, that is not only adequate but is uncontaminated by persistent toxic chemicals. The habitat should contain sufficient shelter, in the form of bankside vegetation, crevices etc., to provide both resting and breeding sites. Additionally, a sympathetic public and enforceable legislation are advantageous. These will form the cornerstones of any conservation strategy.

The first stage in the development of a conservation strategy is to carry out a survey in the field. A broad survey is required to assess the overall distribution of otter populations, followed up by more detailed surveys in areas of particular interest, where positive conservation action is likely to be taken. In some countries, this may include all waterways containing otters, such has been the decline of the species. It is necessary that a programme of surveillance be developed in these areas, i.e. a regular series of surveys designed to detect change. It will also be essential to describe pollutant loads, fish stocks and habitat resources. Such resource surveys should also be conducted in the area surrounding the current range of otters, if the aim of the conservation measures is to extend the range.

Because of the vulnerable status of most populations of the European otter, the legal protection of the species, and preferably also of its habitat, is necessary, though unfortunately such laws are difficult to enforce. A programme of public enlightenment will be essential, if practical conservation measures are to be taken. These practical measures may involve attempts to reduce any pollution loads entering watercourses that could adversely affect otters. If fisheries have been depleted, re-stocking may be necessary. Considerable effort will also be required to retain bankside vegetation and, where necessary, improve cover, both by liaison work with those responsible for managing rivers and by active management. Reductions in pollution and increases in cover are likely to benefit the river community as a whole. Consideration may also be given to reducing pressures on otters from other users of waterspace, for example fishermen or canoeists, but such restrictions are likely to generate considerable controversy and ill-will and should be avoided unless absolutely essential. The effects of any environmental modifications or recreational zoning designed to benefit otters must be properly monitored. As a last resort, where populations of otters are critically low or absent, it may be necessary to consider re-stocking or re-introducing otters, either by captive breeding or by translocating animals from healthy populations.

Legislation

The otter receives legal protection over much of its European range and in 13 countries this protection is complete, either under hunting or wildlife laws. In a further 10 countries, killing otters is allowed under certain conditions, for example at fish ponds or farms.

In Britain, the otter received no legal protection before 1978. However, the Water Act of 1973 stated in Section 22(i) that water authorities shall have regard to the desirability of preserving flora and fauna and, though vague, this has proved a useful lever in protecting some otter habitat. After a long and sometimes heated debate, the otter was addded, on 1 January 1978, to the list of species protected under the Conservation of Wild Creatures and Wild Plants Act 1975, making it an offence to kill, injure or take an otter, or attempt to do these things. This Act applied to England and Wales only and the important populations in Scotland and Northern Ireland remained unprotected. The 1975 Act was superseded by the Wildlife and Countryside Act 1981, which extended protection to Scotland, while similar legislation has recently been passed in Northern Ireland. It is now illegal to kill, injure, take or sell an otter. It is also illegal to disturb an otter in its place of shelter, unless this is done unintentionally, or to damage or destroy any structure used for the animal's shelter or protection. However, a proviso is made that such damage is not unlawful if it was an incidental result of a lawful operation and could not reasonably have been avoided. Furthermore, provided that a licence has been obtained, it is possible to kill an otter, if it can be shown that the action was necessary to prevent damage to livestock or fisheries. Such licences are issued by the departments of agriculture, not by the Nature Conservancy Council.

The Wildlife and Countryside Act has been controversial since its earliest days and parts of it may be unworkable. The ease with which licences may be issued to kill protected species has been strongly criticized (e.g. Prestt, 1984), as has the level of protection for important habitats (e.g. Adams, 1984).

Is it possible to determine all places used by otters for shelter and do they receive protection under the Act? The radio-telemetry study of Green et al. (1984) has shown that an individual otter may rest in a variety of places. Bankside patches of the alien Japanese knotweed (*Polygonum cuspidatum*) are frequently used by otters, but it is now illegal to encourage this plant under the Act! On rivers holding otters, can all such places, which potentially offer shelter, be considered as protected, provided the relevant authorities have been informed? Or is it necessary to prove definitely that a place is being used by an otter? In the latter case, only if an animal is flushed from a site, or only if it is radio-tracked, can one be certain about resting places. If this level of proof is required, then the Act is clearly unworkable as far as the protection of otter habitat is concerned. Clarification is needed and this part of the Act may need testing in a court of law.

As well as national laws, the otter receives protection, at least in theory, under international agreements. In 1979 the Council of Europe Convention on the Conservation of European Wildlife and Natural Habitats was concluded in Berne. The Berne Convention requires the Contracting States to take appropriate measures to protect certain species, including the otter, and their habitats. Twelve states were applying the Convention by 1983 (Dollinger, 1983).

An earlier convention, The Washington Convention on International Trade in Endangered Species of Wild Fauna and Flora (C.I.T.E.S.) was concluded in 1973 and now covers 81 states. The Convention aims to prevent commercial trade between states in endangered animals and plants, whether alive or dead, and in the recognizable parts or products of these animals or plants. The otter received full protection under C.I.T.E.S. in 1976 at the request of Switzerland, after it had been noticed that large numbers of live otters and otter skins from a country in Eastern Europe were being offered for sale.

Education

Because otters can have such large home-ranges, they cannot usually be maintained solely within nature reserves and, although the national parks of continental Europe are large and protected, they tend to be situated in mountainous areas, where the habitat is only marginal for otters. An enlightened public is therefore important for the success of an otter conservation programme. The very nature of the otter makes it an animal popular with the general public, so that the species can be used as a mascot for conservation in general, and wetland conservation in particular (Figs. 1.1 and 1.2). An examination of the 1984–1985 gifts catalogue of the Royal Society for Nature Conservation shows that some 12 separate gift items depict otters, surpassed only by the badger, which is the emblem of the R.S.N.C.

Newspapers and television reach a vastly larger audience and are of great importance in promoting wildlife conservation. The popularity of wildlife programmes with the British public can be judged by the decision of the B.B.C. to show Hugh Miles' highly acclaimed film, *The Track of the Wild Otter*, during peak viewing time on Christmas Day

1983 (see also Miles, 1984). The recent release of captive bred otters (see below) resulted in eight minutes of coverage on a B.B.C. television news programme, while it was widely reported in the media.

The primary role of zoos is for entertainment (as indeed is that of television), but they can also have an important part to play in wildlife education. Their success will depend on the animals being displayed in an interesting environment, along with imaginative information boards. The pitiful row of concrete prisons that frequently passes as a zoo does little more than arouse mild curiosity in bored holidaymakers. A survey by Reuther (1981) found that the size of otter enclosures in zoos ranged from 7 square metres to 15 400 square metres.

In Britain, the Otter Trust is a registered charity concerned entirely with otters. It was established in 1973 under the directorship of Philip Wayre. Its aims are to promote the conservation of otters whenever and wherever necessary for their survival. To do this it promotes and supports field studies and there is a conservation officer on the staff. A collection of otters is maintained at the Trust's headquarters, near Bungay in Suffolk. The grounds are open to the public. In addition to the otters there is also an educational display and a lecture room which is much used by parties of schoolchildren. But the Trust is more than a zoo with a strong educational input. It is notably successful in breeding otters and has released some of these animals to the wild (see below), fulfilling one of its initial aims.

In West Germany, the Aktion Fischotterschutz was established as a registered charity in 1979, under the direction of Claus Reuther. It aims to protect the otter from extinction by maintaining and improving existing otter habitats and by reconstituting damaged otter biotopes. It also carries out research and for this purpose a large enclosure (0·24 ha), together with a range of environmental and behavioural sensors linked to a computer, has been established in the Harz Mountains (Fig. 5.1). Aktion Fischotterschutz also publishes a quarterly magazine, *Otter-Post*, and is very active in promoting the cause of the otter in the media. Its activities are described in more detail by Reuther (1984).

Large enclosures for the study and breeding of otters have also been established in Sweden at the Broda Wildlife Research Station (Sandegren, 1981).

Many countries do not have the resources to establish such facilities, but they are beginning to form otter groups that can mobilize available manpower efficiently to conduct field surveys and develop conservation measures appropriate to their regions. The French Otter Group was the

Fig. 5.1. Aktion Fischotterschutz research pens at Oderhaus, in the Harz Mountains of West Germany.

first in the field. The Gruppo Lontra Italia was established in 1982 under the energetic leadership of Fabio Cassola of the Italian World Wildlife Fund. It has been extremely successful in getting frequent media coverage in a country not renowned for its interest in wildlife conservation and this active group has now embarked on a countrywide survey. Similar otter groups have since been formed between Spain and Portugal, in Belgium and Holland, and in Greece and Switzerland.

Fisheries and pollution

The fish biomass or production in a river required to support a thriving otter population is not known, though it was suggested on p. 23 that oligotrophic streams at high altitudes may have insufficient food to support resident otters. Most other stretches of water should theoretically have sufficient fish to hold otters, and the widespread distribution of otters in Britain during the first half of this century lends support to this view. However, the fish carrying capacity of rivers may be lowered by pollution and habitat modification.

In Britain, information on the species of fish inhabiting a stretch of water, and often some idea of their performance, can usually be obtained from the fisheries biologists of the regional water authorities (or river purification boards in Scotland). For some stretches there may be detailed information. The Anglian Water Authority, for example, plans to have biomass data for the fisheries of all of its 6553 km of watercourse by the end of 1985 (Ronald Linfield, pers. comm.)

Generally, chronic pollution, which results in permanently reduced fish populations, tends to occur in industrial areas or areas with large urban populations, where a variety of factors may mitigate against otters, though exceptions exist (e.g. Fig. 4.1). The increasing and insidious acidification of streams remote from centres of population may, by eliminating fish populations, render large areas of otherwise suitable habitat unusable by otters, though the connection between increased acidification and decreased otter usage is only now being made (authors' unpublished data).

Many spillages of pollutants occur, either by accident or through ignorance. In the region administered by the Anglian Water Authority, for example, almost 70 000 fish were lost in the financial year 1983–1984 in 165 fish mortality incidents, including one on the river where the first release of captive bred otters took place.

Where large fish kills occur, re-stocking is normally carried out by the water authority, who have a statutory duty to maintain fisheries. Fish are also highly mobile and will recolonize from adjacent, unaffected stretches.

In several instances, conservationists have stocked pools or ponds for otters, but inadequate follow up has made it impossible to assess whether otters benefited. Indeed it was not even certain that such stocking was necessary. Before stocking takes place, it is essential to determine what fish are already present. Clearly, overstocking ponds with fish will result in poor quality specimens, while the stock will need to be periodically assessed to see if topping up is required. If stocking was thought necessary, a system similar to that being developed by the Anglian Water Authority, namely off-river supplementation units (Anglian Water Authority, 1984) would be more effective than pond stocking. Carefully managed bankside fish-rearing ponds provide young fish to continuously supplement the river. This would result in general stock enhancement through the range of an otter, rather than concentrating food at a specific and potentially vulnerable site. On the whole, however, if depleted fish numbers are considered to be limiting usage of a river by otters, stocking would clearly be unwise unless the reasons for poor fish performance have been identified and can be effectively ameliorated.

Of much more general concern is the contamination of fish with pollutants. On p. 92 it was described how the sharp decreases in otter numbers were linked to chemicals that are taken up in the food and accumulate in otters. Dieldrin and polychlorinated biphenyls were particularly implicated. In recent years, the usage of these compounds has been much reduced and other wildlife species, notably birds of prey, have shown welcome recoveries in population. Unfortunately, rivers act as sinks for waste products. Many pollutants drain from land into watercourses, which also receive a wide range of industrial effluents. Dieldrin for example is largely restricted in agriculture, but still has a number of industrial uses. We should expect, therefore, that the aquatic environment may still pose pollution problems even though the surrounding land may appear clean.

Fish may contain a wide range of pollutants that are potentially damaging to top carnivores, while themselves appearing in prime condition. Furthermore, contaminated fish may be more vulnerable to predation, thus enhancing the degree of contamination of the carnivore. As an example, fish exposed to cadmium and mercury were more readily caught by predatory fish than were controls (Sullivan et al., 1978; Kania and O'Hara, 1974). Treated fish showed disorientated behaviour within schools and schooling is generally considered to be an anti-predator device.

We should also remember that there is no detailed knowledge of the toxicity of any compound to otters and the species could feasibly be particularly susceptible to a pollutant we have not yet considered important. The case of carbophenothion and geese should serve as a warning. In the early 1970s, large numbers of grey lag and pink-footed geese (*Anser anser* and *A. brachyrhynchos*) were found dead in newly sown cereal fields that had been treated with the organophosphorus insecticide carbophenothion. These two goose species were especially susceptible to this compound, whereas the closely related Canada goose (*Branta canadensis*), as well as domestic chickens and pigeons, were completely unaffected by doses sufficient to kill grey geese (Westlake, Bunyan and Stanley, 1978).

An assessment of pollution risk should therefore be a priority in any programme of otter conservation. The use of fish in this process has already been discussed (p. 101; see also Mason, 1985). There seems little point in spending money and effort on such measures as tree-planting, fencing or re-introducing otters, if contamination with pollutants could be a limiting factor. Pollution assessment should therefore be a first step, though unfortunately this is most often ignored. It should also be stressed that, if it is planned to stock fish to improve food supplies for otters, a sample should be analysed first for pollutant loads; fish suppliers obtain their stock from a wide variety of sources.

Habitat improvement

Once it is clear that food supplies are adequate and uncontaminated, efforts may be made to retain and improve riparian habitat. As Melquist (1984) stated, 'the importance of riparian vegetation to all otters cannot be over-emphasized'. It is too simplistic, however, to blame 'habitat destruction' for an absence of otters. On rivers, we require more precise knowledge of the exact features of bankside vegetation that are important to the animals. Some attempts have been made to discover these (e.g. Macdonald and Mason, 1983c; Bas et al., 1984), but we still do not know the point at which a lack of cover becomes limiting, not only to survival but to breeding success. There are few descriptions of holts used for breeding, features that should be given maximum protection. Some otter resting sites can be located by simple observation, but many more will be overlooked. Therefore, on the basis of descriptions of holts found either by observation or by radio-telemetry, habitat likely to provide suitable sites should, wherever possible, be retained.

In Britain, the Otter Haven Project of the Vincent Wildlife Trust has been promoting the protection of riparian habitat since 1977. Its efforts are centred upon close liaison with water resource managers (King and Potter, 1980) and on the informal co-operation of individual landowners. Most river banks in Britain are privately owned, but the main responsibility for land drainage and flood prevention lies with the regional water authorities. Since it is authority works that alter riverine habitats most frequently and potentially to the greatest degree, close and amicable contact with them is essential. The aim of the Otter Haven Project is to identify rivers where otters still occur, to locate those stretches most frequently used (marked) by the animals and then to establish havens as areas where bankside cover will be retained or improved and human disturbance reduced. Where habitat is considered sub-optimal, trees and scrub may be planted or fences erected to exclude stock and encourage natural regeneration. Ponds close to rivers may be re-stocked with fish. Top priority, however, is given to the protection of known holt sites.

Such aims are clearly laudable, but to achieve results on a meaningful scale large demands must be made on financial and man-power resources. A sympathetic public is also essential to its success.

Many of the havens set up by the Otter Haven Project comprise river lengths of no more than a few hundred metres and are designed to offer protection to one or more known holt sites. David Jenkins (1981), having studied otters on the River Dee system in north-east Scotland, envisaged the otter haven on a more ambitious scale. He thought that a haven should provide breeding habitat, rearing areas (used by the young once they have left the natal holt at about three months old) and habitat for non-breeding animals. To facilitate breeding, he would include small tributaries with their associated woodland and scree slopes to within 500 m of the bank. Jenkins believes that most otters are born away from the main rivers where risks of flooding are reduced. Within his study area, families of young otters were reared on lochs with secure reed beds and woodland and he would include these areas within his potential haven. Habitat for non-breeding otters would be provided by the main river. Thus Jenkins envisages a haven to include about 33 km of river and tributaries together with two lochs, comprising 124 ha. This area, he thinks, should provide security for two to four families of otters. Such a scheme may be possible on the edge of the Scottish Highlands but, however desirable, might be more difficult to achieve in agricultural lowlands. Landowners might well be reluctant to have controls imposed on such large areas.

Jenkins' programme of maintenance for the haven involves retaining present vegetation levels with restrictions on grazing on islands. He advocates the felling of riparian forest in rotational blocks with rapid re-planting. He would place some restrictions on canoeing and confine footpaths to areas away from the river bank (but see p. 149). Water quality should be monitored regularly.

Wayre (1979b), in his account of the havening programme of the Otter Trust also emphasizes the importance of suitable cover and a lack of disturbance as key factors for a successful haven. His aim has been to establish havens along almost entire lengths of river within a restricted region.

In many ways conservation of otter habitat *should* be tackled on the scale of whole river catchments rather than at individual isolated sites. An example of what can be achieved as a first stage can be seen in Lyn Jenkins's report on the Western Cleddau catchment in Wales (Jenkins, 1982). In six weeks she surveyed 96 km of watercourse, including the main river and 13 tributaries. She mapped all sprainting sites (as an indicator of otter activity) and made notes on habitat. As a result she was able to illustrate the relative importance to otters of each part of the catchment. She determined 23 key sites which offered secure resting places for otters and identified three of the tributaries as priority conservation areas. Those stretches of river with evidence of high otter activity had wooded banks, *Rhododendron* thickets or wild *Carex* marshes.

Given this type of background information, the otter conservationist should know which stretches must be, if possible, totally protected from alteration and in which areas it would be beneficial to arouse the sympathy of landowners. He or she should be able to recognize those stretches where habitat improvement might be worth while.

One slight criticism of this project is that the catchment was surveyed once only and it is known that otters can vary their use of different areas. This could be important in terms of any planned habitat improvement. On the River Severn, for example, a summer visit to a bleak and almost spraintless bank might prompt the conservationist to plant scrub. A visit to the same site in winter would reveal conspicuous signal sites (described on p. 48) and, since we do not understand the significance of these sites to otters, it might be better if the bank was left unaltered (Fig. 5.2).

The type of survey undertaken by Lyn Jenkins, an ecologist with knowledge of otter biology, is of undoubted value. A rather more sophisticated approach was attempted in the multi-disciplinary survey of the River Teifi in Wales (Brooker, 1982, 1983, 1984). In theory, the development of an evaluation methodology for assessing wildlife values along rivers is important in providing information on the effects of current river management practices and in formulating ideas for the reduction of deleterious effects. It should also be possible to identify particular river reaches of high wildlife value. The shortfalls in the attempt made on the Teifi are discussed by Eckstein (1984). In this study, analyses of the relationships between indices of otter activity (measured by numbers of spraints, sprainting sites and possible holts) and 26 habitat variables (collected by biologists with no training in otter biology) produced few significant correlations. This was in part due to the large range size of individual otters with regard to specific features of habitat, but there was also a lack of adequate detail on the habitat features recorded. For example, a record was made of those stretches bordered by single lines of trees but no note was made of the tree species. A single

Fig. 5.2. Otters have scratched away the snow at a signal site on a bare river bank.

line of alder may provide a poor resource for an otter, while a line of mature sycamore and oak could offer several safe refuges.

Some recent studies have attempted to link otter activity (measured by marking levels) to specific habitat features (Jenkins and Burrows, 1980; Jenkins, 1982; Macdonald and Mason, 1983c, 1985) and as more people study the animals, we should increasingly recognize the potential importance of individual features. Further results from radio-tracking would be invaluable. Brooker (1982) suggested that, if habitat/resource information could be refined, i.e. the habitat requirements of a species could be identified, then habitat surveys might act as 'surrogates' for the time-consuming species surveys. It may be that, in areas where otters still occur, we are coming closer to that objective.

While habitat conservation on a catchment basis may be the ideal, there is always the temptation, given local encouragement, to try to improve odd corners for otters. Whether this is of real value is open to doubt. Certainly in Britain today more and more farmers are utilizing all grazing land. As a result, on river banks, as mature standards die few

Fig. 5.3. Fencing to exclude cattle and sheep results in rapid herbaceous growth and emergent aquatic plants flourish.

young trees replace them. Figure 5.3 shows the results of fencing to prevent stock from grazing a tiny portion of river bank. After two years there is dense herbaceous growth and a few trees have been planted. It may be that, at some time, an otter will use this cover as a resting site but will the odd patch of suitable habitat ensure that the river is still favoured by otters in 60 years' time? Is there any real value in random aesthetic gestures in the face of continuing destruction of long-established habitat?

In this context, probably the most important work of a conservation organization involves liaison with the regional water authorities. In Britain the authorities have been responsible for the majority of large-scale alterations to riverine habitats but they now have a statutory duty to observe the interests of wildlife conservation. They cannot, however, be held responsible for the destruction of an otter holt if they do not know where it is. The authorities now inform conservation organizations of their plans for capital and maintenance works on rivers so that important otter sites can be identified before the work begins. If the authority has been informed of the location of a holt or a resting place and subsequently destroys it, it is liable to prosecution under the Wildlife and Countryside Act 1981. Many of the authorities are now genuinely sympathetic to the cause of the otter and will try to avoid disturbing river reaches frequently used by the animals. Occasionally mistakes have been made and key sites have recently been damaged but, overall, organizations like the Otter Haven Project have managed to reduce destruction of habitat on many rivers.

It is encouraging also to see some of the authorities replacing or creating habitat for wildlife. On the lower reaches of the River Vyrnwy, for example, where past works have rendered the river a featureless drain, albeit to control severe flooding, the Severn–Trent Water Authority is now planting trees and scrub in a positive effort to increase cover for otters. Many of the other ways in which authority schemes can be adapted sympathetically without reducing their effectiveness are outlined in a comprehensive handbook on rivers and wildlife by Lewis and Williams (1984). Ideally, all budgets for authority capital and maintenance works should automatically include funding for habitat protection and creation.

Tree and scrub planting is, however, a long-term and expensive process and should always be secondary to conservation of existing woody vegetation. With riparian, as with most habitats, the larger the site retained the more likely it is that individual species will continue to survive.

In the past, native plant growth was encouraged on rivers to stabilize banks and to help deflect water flows but as Newson (1984) points out, modern technology and machinery have tempted water managers into abandoning their roots. Between 1930 and 1980 the regional water authorities carried out major or capital works on 8504 km of British rivers and maintained a further 35 500 km (Brookes, Gregory and Dawson, 1983). Much of this work has resulted in the destruction or gross modification of riparian habitats and the removal of marginal vegetation. Yet the arguments for retention of plant growth are convincing, not only in the wide context of preservation of riverine communities but in terms of economic benefit to water resource managers (Table 5.1, and Mason et al., 1984). A well-vegetated bank is a requirement for a properly functioning stream-riparian system (Jackson and van Haveren, 1984).

Studies have shown clearly that more species of birds are found on unmanaged than on managed stretches of river (Royal Society for the Protection of Birds, 1978a,b, 1979) and that some species take years to recover from the effects of river maintenance work (Williamson, 1971). Riparian vegetation also benefits the freshwater community. Invertebrates falling from trees provide food for fishes such as salmonids (Mason and Macdonald, 1982), while the benthic invertebrate community is largely

Table 5.1. *Some values of riparian trees*

Root systems provide holts and resting sites for otters.
Birds find breeding and feeding sites.
Bank erosion is reduced.
Shading reduces aquatic weed growth.
Shading helps to maintain water temperatures suited to salmonids.
Instream roots provide habitats for invertebrates.
Instream roots provide shelter for fishes.
Instream roots create pools for fishes.
Litter input provides food for heterotrophic food chains.
Input of terrestrial invertebrates provides food for fishes.
The landscape is enhanced.

supported by leaf detritus (Cummins, 1979; Kaushik and Hynes, 1971). Shading by bankside trees helps to maintain temperature and oxygen levels suited to salmonids (Meehan *et al.*, 1977; Karr and Schlosser, 1978), while protruding roots can provide much of the instream shelter so important to fish populations (Swales, 1982). The replanting of riparian vegetation has been recommended to benefit fisheries (Kennedy *et al.*, 1983).

Shading also limits macrophyte production (Dawson and Haslam, 1983) and resultant savings in costs of mechanical or chemical weed control can be substantial (Krause, 1977). Thus, in efforts to protect bankside habitat for otters, conservationists can back their case with broadly based supporting evidence.

In many parts of Europe where water resource managers and farmers have already destroyed or devalued much of the riparian habitat, conservationists are acutely aware of the ensuing problems. In other countries, like Greece, where much dense bankside vegetation is still intact, there is a tendency for biologists to be complacent. Farmers, they think, are unlikely to remove the tangles of *Rubus*, *Salix* and vines from river banks because they see no obvious reason for them to do so. In Britain we might have thought the same 30 years ago, but attractive financial incentives to increase productivity can change farming attitudes with some rapidity. Greece has not long been a member of the E.E.C. and Spain and Portugal gain entry in 1986. Efforts should be made now in those countries to ensure that their river banks do not end up resembling the bleak, featureless ones of central England. Yet, the chances are high that little effort will be made and, as usual, conservationists will gain insight from hindsight.

It cannot be stressed too often that retention of long-established cover should always take precedence over attempts at habitat improvement. In Britain, in areas where improvement seems necessary, trees such as ash and oak have been planted close to water in the hope that, as they mature, their roots will erode to provide potential holts. Thickets of willow, thorn and *Rubus* are planted to provide cover more quickly and to offer shelter away from river banks. Attempts have also been made at constructing artificial holts.

Single or multi-chambered dens have been made of brick or stone with pipes providing entrances above or below water level. Some designs are illustrated in a report by the Society for the Promotion of Nature Conservation (undated) and in Wood (1978–1979). Occasionally, water authority engineers have been persuaded to incorporate chambers into stonework used to prevent bank erosion, but a common problem with many constructions has been the silting of entrances. No definite design can be recommended.

In Britain log piles are now being made as potential resting sites for otters. The animals are known to use heaps of flood debris on river banks or piles of discarded wood and brashings left after forestry works. On the River Severn we tracked an otter to a loose pile of straw and debris which had been deposited against a fence during floods one month earlier.

Artificial mounds of wood should be made as large as possible. Natural piles known to be used by otters vary in size from 40 m long, 10 m wide and 4 m high to 5 m × 3 m × 4 m. Clearly these should not be taken as upper or lower limits, but it may be advisable to build on a large scale where possible. No precise plan can be recommended. If large logs or stumps are used for the base, a series of rough chambers of various sizes can be formed. Thinner wood can be used to roof the chambers and debris can be piled on top to make the structure weatherproof. Sites should be chosen above normal high water levels and, if there is any risk of occasional flooding, piles can be wired into place. It would be better to avoid sites that are subject to regular human disturbance and *Rubus* or *Salix* scrub can be planted to make the site more secure and less visible.

The construction of such artificial resting places is recommended tentatively for, while there is little current evidence that otters use them regularly, results of radio-tracking do indicate the use of natural log piles. The construction of artificial sites, however, can never substitute for the retention of natural cover. Jenkins (1981) warns that otters are unlikely

Fig. 5.4. A lake in central Ireland subject to intensive recreational pressure during weekends and holidays. The spit of stones in the background was heavily marked with fresh otter spraints.

to breed in artificial dens, but they could be of value as initial staging posts in areas where long stretches of riparian vegetation are being improved.

Restrictions on river users

It is frequently stated that disturbance is detrimental to otters and the activities of other river users should be controlled. For example, the Joint Otter Group (1977) suggested that 'organizations involved in recreation on waterways should be prepared to accept some limitation on their activities where these might have an adverse effect on otter populations'. King and Potter (1980) state that 'game fishing and wildlife are usually compatible, whereas water-skiing and otters obviously are not'. Jenkins (1981), in his proposals for otter conservation, argues that 'access by the public, including canoeing, should be restricted..., mainly by making it difficult to walk along parts of the river or gain access to them'. Such measures would all restrict otherwise legitimate users of water space and at a time when the requirement for recreation in the countryside is growing. Is there any evidence that such activities are detrimental to otters?

The significance of disturbance has already been discussed (p. 125) and it was concluded that, provided cover is adequate, otters appear to have a high tolerance of human disturbance (Fig. 5.4). Most recreation takes place during the day, whereas otters are now generally active at night. Because restrictions on other users are likely to prove controversial, they should be avoided, unless strong evidence can be provided to show that recreational activities are detrimental to otters.

There are two activities which may directly affect otters – commercial eel fishing and mink hunting. Although it is not possible to determine what effect fyke nets have on otter populations, and most drownings may never come to light, the accumulating evidence suggests that the problem is serious (p. 124) and that, for some endangered populations, fyke nets could result in local extinction.

Conservationists have not attempted to ban eel fishing, but rather to modify the entrances to nets, so that otters cannot get in (Jefferies et al., 1984). These suggestions have caused an outcry from the commercial eel fishermen, who claimed that such modifications would seriously reduce catches and put them out of business. Their general views are presented by Harrison (1984). The outcome of these arguments is that trials are being conducted into the effects of various modifications to fykes on eel catches, with the co-operation of both sides – although only the conservationists are contributing financially!

With the cessation of otter hunting in 1981 many hunts turned to the mink as quarry. In some cases the same individuals and the same hounds are involved, but a number of new hunts have also appeared. At the outset it was considered that mink hunts could disturb endangered otters and the British Field Sports Society approved guidelines for hunting, which included avoiding locations where otters were living (see King and Potter, 1980, for details). These guidelines are certainly not adhered to. Our study area in Shropshire is regularly hunted, including the stretch favoured by the last remaining otters. We even know of one otter haven being considerably shortened to accommodate a mink hunt in the interests of maintaining good will! Many otter hunters remain bitter about the loss of their pastime and resent any control of their new activity.

It is difficult to assess the exact effect of mink hunting on otters. The movement of radio-tracked otters during and following hunts would need to be compared with pre-hunt movements to make a realistic appraisal. Green et al. (1984) concluded their radio-telemetry paper with the consideration that, because mink and otter can share resting places, accidental encounters between otters and hounds seem particularly likely. Thus, intuitively, one feels that mink hunting is likely to cause considerable disturbance to otters.

Can mink hunting be legally prevented on rivers holding otters?

Again, it depends on the interpretation of the Wildlife and Countryside Act (p. 135). If a high likelihood of disturbing an otter in its resting place is sufficient cause to prohibit certain activities, then mink hunting would clearly be untenable on rivers holding otters. If it is necessary to prove that an otter is in residence at a particular site before disturbance becomes illegal, then the law clearly cannot control mink-hunting. Indeed, even if hounds killed a female and her cubs within a holt, the huntsman could avoid conviction by claiming that, however unfortunate, the deaths were an accidental result of a lawful operation and could not reasonably have been avoided. Clarification is required from the legislators, but has thus far not been forthcoming.

Captive breeding

There are a number of pros and cons concerning the breeding of endangered species in captivity and these are critically discussed by Frankel and Soulé (1981). In general, these authors consider that many such captive breeding programmes are ill conceived and they recommend the licensing, by national or international bodies, of such schemes, which should meet certain minimum scientific standards.

One aim of breeding otters in captivity is to supply animals to zoos and wildlife parks, thus avoiding otters being taken from the wild. The more zoos that have otters, the greater the audience that will become familiar with them and their plight, leading to a greater sympathy for their conservation in the wild. The other main reason for captive breeding is for re-introduction of otters to the wild.

In the United States, a Captive Otter Management Project (C.O.M.P.) has been formed to catalogue the status of otters in captivity and to maximize their breeding potential through co-operative studies (Foster-Turley, Davis and Wright, 1983). Otters have generally bred very poorly in zoos. Even such commonly kept species as the Canadian otter and Oriental small-clawed otter breed in only a few zoos throughout the world.

For the European otter, a studbook has been opened and is under the direction of Klaus Robin of the Tierpark Dahlhölzi in Bern. A studbook is an international register which lists and records all captive individuals of a species that is rare or endangered in the wild. Its main purpose is to facilitate the continued and co-ordinated breeding management of the species.

As with other species, the European otter has proved very unwilling to breed in captivity, but considerable success has been achieved by Philip Wayre at the Otter Trust. Over the period 1976–1983, 53 otters have been reared to independence from 28 litters. Female otters seem especially sensitive to disturbance at parturition and in the first few days after birth. Much of the success of the Otter Trust may be due to the fact that the otters are left completely alone during this period.

The Alpenzoo at Innsbruck has also had some recent successes in rearing otters (see Fig. 2.18*b*,*c*,*d*).

Re-introduction

One of the main aims of a captive breeding programme is to re-introduce or replenish stocks in the wild. However, the area is a controversial one, with both strong advocates and people equally strongly against. We need to begin by defining some of the terms and follow those given in Boitani (1976):

Introduction: the deliberate or accidental release of animals or plants of a species into an area in which it has not occurred in historical times; or, a species or race so released.

Re-introduction: the deliberate or accidental release of a species, or race, into an area in which it was indigenous in historical times; or, a species or race so released.

Re-stocking: the deliberate or accidental release of a species or race into an area in which it is already present.

In practice, as far as otters are concerned, we are only interested in re-introduction or re-stocking. A working group (Anonymous, 1979) suggested a number of criteria that should be considered before re-introduction was allowed in Britain. First of all, there has to be a clear understanding of why the species was lost. Only those lost through human agency and unlikely to recolonize naturally are considered suitable candidates for re-introduction. There should be suitable habitat of sufficient extent and isolation to which the species can be re-introduced. The animals taken for re-introduction should be genetically as close as possible to the native population and their loss should not prejudice the survival of the population from which they are taken.

There is no doubt that man and his activities have been responsible for the loss of otters. However, whether one should re-introduce, or await natural recolonization, is a contentious issue. In Britain there are still good populations in the west and north and in theory, if conditions are suitable, recolonization should take place. However, whereas the extinction of otters over a wide area took only a few years, recolonization, even in ideal conditions, would take many decades.

There are few nature reserves in western Europe of sufficient size to retain a viable population of otters, so re-introduced animals will have to live in an environment that may house many threats. The local populace must therefore be sympathetic to the presence of otters on their rivers. Even then the disenchanted hunter or fisherman could quietly frustrate a re-introduction programme, especially as only small numbers of animals are likely to be involved.

Pollution and habitat have already been discussed in this chapter. It is clearly essential to ensure that there is sufficient food, that it is uncontaminated and that there is sufficient habitat for resting and breeding before a re-introduction programme is enacted.

There are behavioural and genetic problems in releasing captive bred animals to the wild. Behavioural changes may occur in captive populations that may affect the success of re-introductions. These include an inability to mate or rear young, an inability to forage or escape predators, and a loss of fear of man (Kleiman, 1980). Genetic variability may be lost from small, captive bred populations (Chesser, Smith and Brisbin, 1980). Small populations lose genetic diversity through inbreeding and through random drift of gene frequencies due to sampling variance from generation to generation (Frankel and Soulé, 1981; Ralls et al., 1983). These processes may merge into one in small populations and the results lead to increased mortality in young animals and reduced fertility in adults. Such deleterious effects have been recorded in a wide range of species, including mink (Johansson, 1961).

All re-introductions of otters from captive-bred stock are likely to be based on very small numbers of animals. The basic rule of conservation genetics, according to Frankel and Soulé (1981), states that the maximum tolerable rate of inbreeding is 1%, which translates into an effective population size of 50. Myers (1979) succinctly states that 'a vertebrate breeding stock with fewer than 50 individuals is liable to carry a built-in potential for its own destruction, since in-breeding brings together the harmful genes that larger pools can accommodate'. Berry (1983) takes a more optimistic view, arguing that genetic architecture protects variation and reduces the rate of loss in small populations.

An early attempt at re-stocking with otters was made in Dumfries-shire,

Scotland, where animals from Norway were released for hunting, though no records were kept (Stephens, 1957). A pair of otters, obtained from a Swiss zoo, were released in Haute-Savoie, France, in April 1972, and were apparently still surviving in 1980 (Kempf, 1981), although it seems that they did not breed.

By 1975, so few otters remained in Switzerland (Müller *et al.*, 1976), that re-introduction was considered. Several areas were examined and fish stocks, habitat and potential threats were assessed, though apparently this did not include the analysis of fish tissues for pesticides. Four male and four female otters were procured in Bulgaria and released in the Schwarzwasser/Sense area in the canton of Berne. Two of the otters were found dead within a year, but a population became established and appears to be reproducing, as evidence of juveniles was found in both 1981 and 1982 (H. U. Müller, unpublished data, *per* Raymond Lebeau).

One of the main reasons for setting up the Otter Trust was to provide captive-bred animals for re-introduction (p. 137) and this objective was achieved in 1983. In eastern England, a few small populations of otters remain, which may be too isolated to make contact with one another. If captive bred animals are released into the gaps between wild populations, they could reproduce so that populations eventually coalesce (Jefferies and Mitchell-Jones, 1981; Jefferies and Wayre, 1983). For the first release, three young otters, two females and one male, were selected. They were maintained in a large enclosure, away from the public so that they did not become accustomed to humans. It was planned to release them when aged about 18 months, when they had passed the period of greatest mortality for otters and were approaching sexual maturity.

A number of potential sites for release were examined and on the river chosen, the riparian landowners were both enthusiastic and co-operative. The habitat was considered adequate, as was the water quality and fish biomass. To assess the level of contamination in fish, five chub (*Leuciscus cephalus*) were caught and analysed for organochlorines (certainly a minimum requirement for such a study). Low concentrations of lindane, DDE and PCBs were found, but were considered unlikely to be of significance.

Twenty days prior to release the three otters were placed in a small pre-release pen ($9 \cdot 1$ m $\times$ $15 \cdot 2$ m) on an island by the river. The pre-release pen was to ensure that the otters became used to the sounds and smells of the river. Six days before release the otters were examined and weighed, while the male was fitted with a harness and radio-transmitter. The release took place at dusk on 5 July 1983. During the first night the otters remained in thick cover within 30 m of the release pen and they returned for seven nights for food provided for them. The radio-transmitter remained on the male for 50 days following release. During this time the range gradually extended to 15·3 km of river, while over the next 50 days, the distribution of spraints indicated that a minimum of 31·5 km were in use (Jefferies, Jessop and Mitchell-Jones, 1983; Jefferies, 1984). The prints of a cub, following those of a female, were observed in August 1984, showing that breeding had occurred on the river.

During 1984, the Otter Trust released five more otters onto rivers in East Anglia, while more releases are planned during 1985.

If successful breeding is achieved, and no problems relating to habitat quality or contamination of food occur, the release of captive bred animals will make a useful contribution to enhancing otter populations in East Anglia. However, the process is necessarily slow and long term, only a few animals each year being available for release. The genetic diversity may also be limited. To supplement these animals, translocation of otters from healthy populations should also be considered. The authorities in Britain have, thus far, been very guarded about trans-location. Otters have proved difficult to catch and it is suggested that removing animals will deplete donor populations. It is said that otters are adapted to local conditions and may not survive in a new environment. Finally, worries are expressed about the mixing of races.

In North America, translocations of both river otters and sea otters have been very successful and there seems no reason *a priori* why the European otter should be any less adaptable. Populations in some areas of the British Isles, notably Scotland and Ireland, must be at carrying capacity and could withstand the removal of animals for translocation purposes. Indeed, there are increasing complaints about otters at fish farms and licences may well be issued for their removal or elimination. We have heard of plans in Scotland to trap troublesome animals and put them into areas within Scotland where otter numbers are depleted. But such areas should be quickly filled by natural recolonization if they were suitable for otter survival. Their translocation to suitable areas in England would be preferable.

The subspecies problem should be largely irrelevant in Europe, because none is accepted, but objections to translocations have been made because of the possibility of, as yet, unrecognized races existing! Frankel and Soulé (1981), however, argue that most subspecies have little biological meaning, their delineation being largely arbitrary. They point out that hybridizing between subspecies may be beneficial to inbred lines. Thus the introduction of translocated animals may be valuable, especially into populations, such as those in East Anglia, which have been small and isolated for a number of years.

Any programme of release of captive bred animals or of translocation needs to be carefully planned and co-ordinated by a central authority for the greatest effectiveness. In the future, provided that environmental conditions are found to be suitable, a number of countries will express interest in projects of re-introduction or re-stocking and there will be much scope for fruitful co-operation.

Conclusion

In this chapter we have outlined the basic requirements for otter conservation, while admitting that information on many aspects of otter biology is still meagre. We may, however, seem less than enthusiastic about many of the conservation measures that have been taken. Money for conservation is in short supply. When funds are found for a project it is essential that as much information as possible be gleaned, so that similar programmes carried out later, or elsewhere, can benefit from the experience. Projects must be carefully planned and fully *monitored*. Blind faith is a poor substitute for a well executed plan and will ultimately result in an unnecessary waste of resources.

Much of the practical conservation carried out for the otter has been a last-ditch attempt to save the species from local extinction – the fire-brigade response (Myers, 1979). However, there are still countries with widespread and healthy otter populations. In Portugal and Greece (pp. 68 and 73) otters still occur, for example, in irrigation canals in highly agricultural areas. Dense vegetation remains on the river banks and the agriculture has not developed to the stage where ecosystems are widely polluted with persistent pesticides. Yet within the next decade 'improvement' of marginal land such as river banks and increases in the use of biocides could result in a substantial reduction in the otter's range.

Action in Greece and Portugal must be taken *now* to conserve both otters and the general health of river environments. However, in both countries, the numbers of active conservationists are, at present, pitifully few and their problems in gaining governmental and public support immense. We have discussed the problems with government officials, who at the time expressed an interest, but the situation proceeded no further. Two British fieldworkers are unlikely to be experienced in the ways of foreign governments. Obviously, a joint approach with the international conservation agencies is required, where sound biological data can be matched with political skills and prestige. Unfortunately, this crucial type of support is often lacking.

The otter can be seen as a key species in the wetland environment. Because of its sensitivity to both pollution and habitat loss, conservation measures aimed at retaining a healthy otter population will maintain the integrity of the wetland ecosystem. In Greece, for example, this would include the conservation of a number of internationally important wetland areas, such as Mikra Prespa (Fig. 3.17). Indirectly, it may benefit the much maligned and discussed Mediterranean Sea, because pollution inputs would be monitored and controlled. Nevertheless, we sadly predict that, when we repeat our Greek otter survey in 1991, the ensuing decade will have seen a sharp decline in range, with the otter restricted to the uncultivable uplands.

6 THE WORLD'S OTTERS

'Beyond the Wild Wood comes the Wide
World' said the Rat.... 'I've never been
there, and I'm never going, nor you
either, if you've got any sense at all'.

KENNETH GRAHAME,
The Wind in the Willows

The previous chapters have dealt in some detail with the ecology and
conservation of the European otter. In this final chapter we will review
what is known of the ecology and conservation of the other otters of the
world. For most species there is a dearth of information and the situation
is made worse by the confusing taxonomy of the Lutrinae. Harris (1968),
following earlier workers, listed 19 species of otters, with a total of 63
accepted subspecies. The taxonomy of American otters was revised by
van Zyll de Jong (1972), based largely on a statistical analysis of cranial
and dental morphology. He reduced the number of South American
species of *Lutra* to three, whereas Harris listed seven, and placed them
in a new genus *Lontra*, together with the Canadian otter. Corbet and
Hill (1980) listed 13 world species and their classification is followed here
(Table 6.1). However, Davis (1978) considered that previous studies,
limited largely to cranial morphology, did not yield the true relationships
between species because of the high degree of both convergence and
adaptive radiation. He used both morphological and behavioural data to
divide the subfamily Lutrinae into three Tribes (Lutrini, Aonychini and
Hydrictini), containing a total of only nine species. His classification is
also shown in Table 6.1.

Table 6.1. Otters of the world

Scientific name	English name	Head and body length (cm)	Tail length (cm)	Weight (kg)	Davis' classification
Lutra canadensis	Canadian otter	66·0–107·0	31·5–46·0	8·2–11·3	No change
Lutra felina	Marine otter	57·0–78·7	30·0–36·2	3·2–5·8	No change
Lutra longicaudis	Southern river otter	50·0–79·0	37·5–57·0	10·5–14·8	Subspecies of L. canadensis
Lutra lutra	European otter	57·0–70·0	35·0–40·0	4·0–11·4	No change
Lutra maculicollis	Spotted-necked otter	57·5–66·0	33·0–44·5	3·2–8·5	Hydrictis maculicollis
Lutra perspicillata	Smooth-coated otter	66·0–79·0	40·6–50·5	7·3–11·2	Lutrogale perspicillata
Lutra provocax	Chilean otter	57·0–70·0	35·0–46·0	No data	Subspecies of L. canadensis
Lutra sumatrana	Hairy-nosed otter	70·0–82·6	35·0–50·9	5·0–8·0	Subspecies of L. lutra
Pteronura brasiliensis	Giant otter	96·0–123·0	45·0–65·0	24·0–34·2	No change
Aonyx capensis	African clawless otter	72·5–95·0	40·6–57·0	15·8–20·5	No change
Aonyx cinerea	Oriental small-clawed otter	40·6–63·5	24·6–30·4	2·7–5·4	No change
Aonyx congica	Zaïre clawless otter	79·6–95·0	50·0–56·0	13·6–18·1	Subspecies of A. capensis
Enhydra lutris	Sea otter	55·0–136·0	12·5–33·0	23·8–36·4	No change

Note: Measurements are from Harris (1968).

Fig. 6.1. The Canadian otter, *Lutra canadensis*. Note the large rhinarium, or hairless part of the nose. (Photo: Dennis Hada.)

Fig. 6.2. Historical range of *Lutra canadensis*.

Canadian otter or North American river otter

Lutra canadensis (Fig. 6.1) is the North American counterpart of the European otter and is similar in size and general appearance. The most conspicuous difference is in the size and shape of the rhinarium, the hairless part of the nose, which is much larger in *L. canadensis* than in other species.

The Canadian otter ranged originally from arctic Alaska, at a latitude of about 70° N (Magoun and Valkenburg, 1977) to the southern United States of Florida and Texas (Fig. 6.2), inhabiting lakes, streams, coastal saltmarshes and, in some areas, rocky sea coasts. The range, however, has contracted markedly during the present century (Fig. 6.5).

The Canadian otter, like the European otter, is primarily an aquatic feeder, the diet consisting largely of fish. It appears to catch these in proportion to their abundance and inversely in proportion to their swimming ability (Ryder, 1955). The North American fish fauna is much richer than that of Europe and consequently the otters take a greater variety of fish species. The data in Table 6.2 show that salmonids make up a small proportion of the diet in most studies, though Melquist *et al.* (1981), in Idaho, found otters concentrating their attention on spawning runs of kokanee (*Oncorhynchus nerka*), 93% of spraints containing remains of this species in September, when the spent fish begin to die. Mountain whitefish (*Prosopium williamsoni*) and large-scale suckers (*Catostomus macrocheilus*) were also major food fish in Idaho and most individuals taken were more than 30 cm in length. Otters on the Pacific coast of British Columbia ate species of fish similar to those taken by European otters on rocky northern coasts.

As in Europe, the Canadian otter has frequently been destroyed for its supposed depredations on fisheries, but Ryder (1955) considered that otters may benefit fisheries by removing competitive and predatory fish from trout waters, and more recent work has not refuted Ryder's thesis.

Crayfish are important in the diet in some areas and were the major food in a freshwater marsh in California, where they were abundant. Crabs were taken in large numbers by otters living in a Californian coastal lake, but crabs did not feature prominently in the diet of coastal otters in British Columbia, in contrast to the situation in Europe. In California, but nowhere else, freshwater mussels (*Anodonta* sp.) were taken extensively (Morejohn, 1969).

The proportion of other invertebrates reported is variable. Some authors (e.g. Greer, 1955; Knudsen and Hale, 1968) consider they are of moderate importance as food, but the majority of workers conclude that small invertebrates are taken incidentally. Melquist and Hornocker (1983), however, have observed otters foraging for stonefly nymphs and diving beetles, though in terms of biomass, such invertebrate prey must be of little significance.

In most studies, vertebrates other than fish made up only a small proportion of the diet. In Montana, amphibians accounted for more than 10% of the diet, while in other studies they were not reported. Reptiles,

Table 6.2. *Diet (expressed as relative frequency, %) of Lutra canadensis in North America*

Locality	Habitat	Sample size	Method	Salmonidae	Other fish	Crustaceans	Other invertebrates	Amphibians	Reptiles	Birds	Mammals	Vegetation	Reference
Michigan	Various fresh waters	393	G	3·7	74·1	11·7	5·4	4·7	—	tr	tr	—	1
Montana	Lakes/streams	1374	S	11·2	45·4	—	25·1	11·2	0·2	3·1	3·7	—	2
Michigan	Various fresh waters	54	S	2·8	82·9	4·6	3·9	5·7	—	—	2·8	—	3
New York	Various fresh waters	141	G	3·3	44·0	23·4	9·1	16·7	—	0·5	1·8	—	4
Massachusetts	Reservoir	517	S	—	55·7	27·9	12·1	—	—	0·6	1·9	1·5	5
Great Lakes	Lakes/streams	501	G	2·0	54·8	25·5	11·5	4·2	—	—	4·7	—	6
Michigan	Rivers/pools	47	S	1·0	79·0	12·0	1·0	2·3	—	—	—	—	7
Oregon	Various fresh waters	75	G	15·9	62·8	22·1	7·1	8·0	—	5·3	0·6	—	8
Alabama/Georgia	Various fresh waters	315	G	—	49·1	36·9	7·9	3·2	—	0·2	3·6	2·2	9
California	Freshwater marsh	220	S	—	15·9	52·7	7·3	—	0·5	20·5	—	6·8	10
California	Coastal lakes	100	S	—	56·1	37·6	1·3	—	—	4·6	—	—	11
Idaho	Lakes/streams	1902	S	30·1	63·2	—	4·0	—	tr	1·4	1·2	—	12
Massachusetts	Lakes/streams	56	G	5·7	78·5	0·4	tr	15·4	—	—	—	—	13
Alberta	Lakes/streams	498	S	—	64·2	0·7	17·5	0·7	—	11·7	5·2	—	14
British Columbia	Sea coast	528	S	0·2	89·8	6·5	—	—	—	3·8	—	—	15

Note: Methods: S = spraint analysis; G = gut analysis; tr = trace. References are as follows: 1, Lagler and Ostenson (1942); 2, Greer (1955); 3, Ryder (1955); 4, Hamilton (1961); 5, Sheldon and Toll (1964); 6, Knudsen and Hale (1968); 7, Field (1970); 8 Toweill (1974); 9, Luahachinda and Hill (1978); 10, Grenfell (1978); 11, Modafferi and Yocom (1980); 12, Melquist *et al.* (1981); 13, Loranger (1981); 14, Gilbert and Nancekivell (1982); 15, Stenson, Badgero and Fisher (1983).

mostly snakes, were recorded very infrequently, as were mammals. Birds were occasionally important features of the diet. In Suisin Marsh, California, they made up 20·5% of the diet, but here peak waterfowl numbers may reach 700000 (Grenfell, 1978). Birds were also important in Alberta, where large populations of moulting wildfowl occur. Predation on birds is probably largely confined to freshly killed, sick, crippled or young birds (Melquist and Hornocker, 1983). Foottit and Butler (1977) describe otters raiding island colonies of glaucous-winged gulls (*Larus glaucescens*) and taking young birds. Conversely, Kilham (1982*a*) observed crows (*Corvus brachyrhynchus*) stealing food from otters; one crow would peck the tail of the otter, which swung round in defence, allowing another crow to steal the fish!

Most vegetation recorded in the diet of otters must be taken incidentally, but the frequency of blueberries (*Vaccinium* sp.) in one study (Sheldon and Toll, 1964) suggests that otters occasionally take fruit.

When feeding, otters do not generally remain long at one specific location but tend to travel throughout their home range, visiting preferred foraging sites. However, in Idaho, otters congregated at spawning areas of kokanee, eight to twelve otters being recorded at one such place (Melquist *et al.*, 1981). In the south of its range, the otter is influenced by seasonal droughts. In the Everglades of Florida, otters occurred in swamps and marshes during the wet season, but they retreated as marshland dried and were restricted to permanent ponds towards the end of the dry season (Humphrey and Zinn, 1982). During a drought year in Florida, Kilham (1982*b*) found otters regularly visiting a small pool of 8 m × 20 m, dug for watering cattle, and feeding exclusively on walking catfish (*Clarias batrachus*).

The field signs of Canadian otter are very similar to those of European otter (Mowbray, Pursley and Chapman, 1979), but they have been used very little in ecological studies. Otter lure, a commercially available scent for use by trappers, has been placed on boards in wetlands to encourage sprainting by otters, in an attempt to study seasonal habitat use (Humphrey and Zinn, 1982). However, the intensive radio-telemetry study by the University of Idaho has provided most information on home-range, activity and social behaviour of Canadian otters.

The Idaho study area was the catchment of the North Fork Payette River, a system of streams, rivers and lakes in a largely wild, mountainous area. The study was conducted between 1976 and 1981, during which time 39 otters were monitored by radio for a total of 4888 hours during

2656 days on which the otters were located. The scale of this study is thus far grander than anything so far attempted in Europe. The methodology is detailed in Melquist and Hornocker (1979) and the results in Melquist and Hornocker (1983). The researchers initially placed collars on their otters to carry the radio-transmitters, but they experienced continuous problems, so that a transmitter was developed that could be implanted into the peritoneal cavity.

Within the study area as a whole, otters preferred valley to mountain habitats. Food was the main factor influencing distribution, but adequate bankside cover was also important. When both food and cover were excellent, otters could tolerate the proximity of man. There was approximately one breeding female per 20 km waterway, one adult male per 53 km and one yearling or non-breeding adult per 14 km, the overall density being estimated at one otter per 3·9 km waterway. The home-ranges of individuals showed extensive overlap. Seasonal home-range varied in length from 8 to 78 km, the smallest home-range for an animal (an adult female) radio-monitored over an entire year being 31 km. Within the home-range were a number of activity centres, located in areas of abundant food and adequate shelter (cf. European otter, p. 28).

The basic social group consists of an adult female and her young. However, Melquist and Hornocker (1983) considered that otters were more social than other mustelids, with lone animals frequently associating with other otters and family groups, though usually for short periods of time. Cubs stayed with their mother for between 7·5 and 11·5 months. Juveniles dispersed during April and May at an age of 12 to 13 months, though the extent of dispersal was very variable and some animals did not disperse at all. Dispersal occurs before sexual maturity. One dispersing male covered 104 km over 30 days, while a female covered 192 km over 50 days. The dispersal of juveniles ensures that the female is isolated when giving birth and rearing her next litter of cubs.

Melquist and Hornocker (1983) found much variation in the daily movements of individuals. Food availability was probably the main influence on movements. The mean daily distance travelled was generally less than 5 km, but one dispersing yearling male covered 42 km in a day.

Otters were active by day and night (Fig. 6.3), the extent of daily activity probably being due to the wildness of the area and general lack of human influence. Overall, there was more night-time activity, but daytime activity predominated in the winter. During spring and summer, peaks of activity occurred around midnight and dawn, with a slight increase in mid-afternoon, and a further increase towards dusk. During

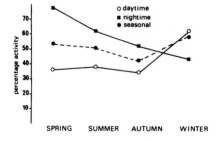

Fig. 6.3. Diurnal, nocturnal and seasonal activity of Canadian otter, based on 6266 hours of activity recordings. (Data from Melquist and Hornocker, 1983.)

the autumn and winter, otters were less rhythmic. Overall, otters showed least activity in autumn and most in the winter (Fig. 6.3). Foraging and feeding occupied the majority of the otters' activity, being recorded during 62% of the observation periods. By contrast, play occurred in only 6% of observations and involved mainly the cubs, contrary to the popular opinion of the otter as a playful animal.

In the Idaho study, otters used a variety of den and resting sites. As with the European otter, they change their resting sites frequently, one animal using 88 sites over a 16 month period. Otters do not excavate their own dens. Beavers' dens were used most often (32% of the total), probably because they were readily available and provided shelter with underwater escape routes. Log-jams were also important, because they offered both shelter and excellent feeding close by.

As with the European otter, scent marking by faeces and urine was the most important method of communication and scats were deposited in places similar to those described for *Lutra lutra* on p. 47. Melquist and Hornocker (1983) considered that such communication enabled otters to exhibit a 'personal space dispersion', whereby an individual and its current location could be defended without reference to fixed spatial boundaries by mutual avoidance between individual otters. Hornocker, Messick and Melquist (1983), however, stress that the behavioural system is very flexible, changing as environments change. Spatial strategies adopted by *L. canadensis* may be different in other populations. Furthermore, although the Idaho population appeared not to be strictly territorial, Hornocker *et al.* consider that populations completely free of human exploitation may exhibit a more traditional territorial system. Studies in other areas, using radio–telemetry, would be extremely interesting.

From the above discussion, it is clear that the Canadian otter is ecologically very similar to the European otter, but they are very different in their reproduction. The Canadian otter exhibits delayed implantation

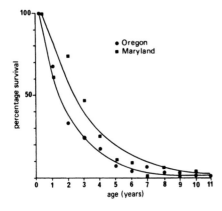

Fig. 6.4. Survivorship curves for female *Lutra canadensis* from Oregon and Maryland. (From data in Tabor and Wight, 1977, and Mowbray *et al.*, 1979.)

(Hamilton and Eadie, 1964), in which the new embryo, or blastocyst, formed after fertilization does not implant on to the wall of the uterus, but remains unattached, without developing, for several months. Implantation is delayed for about 10 months, giving an apparent gestation period of about 12 months. Why there is this difference in reproductive strategy between closely related species has been the subject of much recent discussion, particularly in relation to the stoat (*Mustela erminea*) and weasel (*Mustela nivalis*) (e.g. Sandell, 1984; King, 1984). Considering a number of hypotheses, King favours the view that weasels and stoats, despite their apparent similarity, descended from different distant ancestors, one with and one without delayed implantation. This historical hypothesis may be the most acceptable in the case of the otters and, as mentioned on p. 159, van Zyll de Jong (1972), on morphological evidence, concluded that the two species were not closely related, placing *canadensis* in a new genus, *Lontra*.

Cubs are generally born in late winter and early spring, though Humphrey and Zinn (1982) consider that, in Florida, mating, and hence subsequent births, occur mainly in autumn, to coincide with seasonal flooding. The mean litter size, estimated from foetal counts of killed animals, is reported as 2·29 to 2·75 (Hamilton and Eadie, 1964; Mowbray *et al.*, 1979; Tabor and Wight, 1977), while Melquist and Hornocker (1983) observed litter sizes of between one and four (mean 2·4), similar to those of the European otter. The development of cubs in captivity is described by Johnstone (1978) and weaning takes place at about five months (Liers, 1960). Female otters first breed when two years old.

The structure of two populations has been determined by ageing female carcasses collected from hunters and survivorship curves can be constructed from these (Fig. 6.4). Over 50% of females are dead before

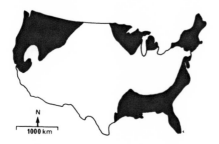

Fig. 6.5. Current distribution of *Lutra canadensis* in the U.S.A.

the age of three. The only comparable estimate of mortality for European otters indicates a similar high mortality in the first two years of life (Stubbe, 1969). Apart from man, a number of predators have been recorded as taking otters (e.g. alligators, bobcats and coyotes), but their impact is likely to be small (Toweill and Tabor, 1982).

The historical range (Fig. 6.2) of the otter was throughout most of North America, excepting the far north and the arid south-west but, from about the middle of the last century, declines began in many areas. The species is now largely absent from the interior of the United States (Fig. 6.5) and from the southern prairie areas of the Canadian provinces of Alberta, Saskatchewan and Manitoba. Undoubtedly, a few animals survive in these regions, for otters have been recorded in the last 10 years in Nebraska and Tennessee (Farney and Jones, 1978; Field, 1978). Jenkins (1983) reported the otter absent in 11 states and rare in a further 13.

Reasons put forward for the decline are habitat destruction, water pollution and unregulated trapping. Undoubtedly, extensive habitat destruction has occurred in the United States during the present century, but there must be many wild areas left in central U.S.A. of sufficient size to support viable otter populations. The extent of wetland and riparian habitat destruction appears not to have been quantified and the use of habitat by otters in a range of habitat types has not been studied. Habitat destruction may be a contributory cause in the demise of the otter in the American interior, but it is unlikely to be the main cause.

The concentrations of bio-accumulating pollutants in Canadian otters have been reported in Chapter 4 (see Tables 4.1, 4.2, 4.4). Considering that such a potential source of material is available from carcasses discarded by trappers, comparatively few studies have been undertaken, and they have tended to examine specific compounds, rather than comprehensively describe pollutant loads. For example, dieldrin has only been reported for two studies, whereas DDT, which is not especially

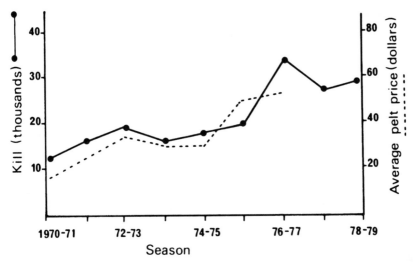

Fig. 6.6. Annual number of otters trapped and average pelt prices in the U.S.A. (From Nilsson, 1980.)

toxic to mammals, has been more widely investigated. Organochlorine pesticides, PCBs and mercury have all been suggested as possible factors in the decline of otter populations. However, no attempt has been made to plot population trends with time and relate these to the introduction of specific compounds, as has been done in Britain by Chanin and Jefferies (1978). If the decline was catastrophic, as occurred in Britain, then trapping returns, analysed with care, could detect such a change.

The otter, however, was extinct in several states before the turn of the century and probably reached a population low around 1935 (Jenkins, 1983), before the main period of use of pesticides. Excessive trapping must have been the cause of this. Contrary to the situation over most of Europe, in North America the otter is classed as a 'fur bearer' and the authorities attempt to manage populations to obtain a sustained yield of pelts. Otters are killed for their fur in 25 states, though they are economically important in only about 14 (Jenkins, 1983). There has been an increase in the number of otters trapped during the 1970s, which can be related to the value of the fur (Fig. 6.6). Some 29 000 otters were killed during 1978–1979 in the U.S.A., with over 9000 from Louisiana alone. In Canada, the number killed was about half that in the U.S.A. (Jenkins, 1983). In addition to the deliberate trapping of otters, the animals are taken as an indirect catch during beaver trapping and, as the majority of traps are indiscriminate, even those states protecting the otter suffer otter mortalities.

Because of the endangered status of otters over much of the world,

L. canadensis was placed in Appendix II of C.I.T.E.S. (Convention on International Trade in Endangered Species of Wild Fauna and Flora) in 1977, giving it potentially endangered status. All otter skins must be tagged before export. The majority are sent to Europe, where, of course, the local otter is protected!

To manage the otter populations adequately for exploitation, population monitoring is required. However, the main source of population data is the trapping returns themselves and these are influenced both by the value of the otter fur and also by changes in the level of beaver trapping, making trends meaningless (Tumlison, King and Johnston, 1981). Additional data are collected from small samples of carcasses, but there seems a great reluctance on the part of the biologists of the regulatory agencies to attempt field surveys, as are done by conservationists in Britain.

In some states where otters have disappeared, re-introductions are being carried out by translocating otters from other states. In Colorado, otters were last recorded in 1921, but they are now being re-introduced from Wisconsin, Washington and Newfoundland. Tennessee and West Virginia also have re-introduction plans (Jenkins, 1983) and American schemes tend to be on a much larger scale than those reported in Britain on p. 154. In addition, the River Otter Fellowship has been formed to increase public knowledge and interest in otters and to give them protected status.

Neotropical *Lutra*

Corbet and Hill (1980) list three species of *Lutra* occurring in Central and South America – *longicaudis*, *provocax* and *felina* – while Davis (1978) considered the first two of these to be subspecies of *canadensis*. The classification of Corbet and Hill is followed here. The distributions of the three species are shown in Fig. 6.7. *Lutra longicaudis* has the widest distribution from northern Mexico (Brown *et al.*, 1982) to northern Argentina, while *L. provocax* is restricted to parts of Chile and Argentina. *Lutra felina* occurs on the coastline of Chile and Peru and in the past Argentina, though it may be extinct now in this country. The ecology of all three species is very poorly known.

Lutra longicaudis lives in a variety of habitats, from small forest streams to coastal savannah swamps. Laidler and Laidler (1983) reported the species living in irrigation ditches among fields of rice and sugar cane

Fig. 6.7. Distribution of *Lutra* in Central
and South America. (From information in
Melquist, 1984.)

in Guyana, giving spatial separation from the giant otter (*Pteronura brasiliensis*), which avoided areas inhabited by man. Duplaix (1980), however, found both species sharing the same creeks in Surinam, at least during the rainy season. *Lutra longicaudis* feeds mainly on fish and crabs, taking smaller specimens than does the giant otter. It appears to mark its home-range with spraints placed at frequent intervals, in places similar to those used by *L. canadensis* and *L. lutra*; a scratched-up sand pile was found by Duplaix (1980). The species is largely crepuscular and nocturnal. Cubs were reported in October and December in Surinam. The rivers of South America contain a number of potential predators; a pet animal was killed by an anaconda (*Eunectes murinus*).

Even less is known of *L. provocax*. It occurs in rivers, lakes and estuaries, overlapping in some areas with *L. felina*. Its distribution lies between the latitudes of 36° S and 52° S. The diet appears to consist almost entirely of crustaceans (*Aegla, Sammastacus*) and bivalves, and introduced salmonid fish appear not to be taken (Chéhebar, 1985). Indeed, salmonids have been implicated, though without foundation, in the decline of the otter, through competition for crustacean food.

The small and distinctive marine otter, *L. felina* lives mainly in the littoral zone, though it sometimes enters rivers to seek freshwater prawns (Hernandez, 1960). It occurs mainly on rocky shores, buffeted by strong winds and waves, and seeks shelter in caves and tunnels above the high tideline in dense vegetation of scrub and small trees (Cabello, 1978, 1983). The habitat appears very similar to that used by coastal otters in western Scotland.

The daytime activity of marine otters has been studied on the Isla de

172 OTTERS: ECOLOGY AND CONSERVATION

Chiloe, in Chile. Otters showed two periods of activity, between 10.00 and 12.00 hours and between 15.00 and 17.00 hours (Cabello, 1983), but nothing is known of their activity at night.

Otters were watched fishing by Cabello (1983) on many occasions. Shellfish (crustaceans and molluscs) comprised 73% of the diet and fish 27%. Spraint analysis showed that the shellfish consisted of 72·2% crustaceans (*Homalaspis plana* and *Taliepus dentatus* predominating) and 27·8% molluscs (especially *Concholepas concholepas*). The spraints also contained large quantities of seeds of the bromeliad *Fascicularia bicolor*, suggesting that they were taken deliberately.

Young marine otters seem to be born mainly in autumn and winter. The litter size is generally two, though four to five have been recorded in some populations (Sielfeld, Venegas and Atalah, 1977; Cabello, 1983).

All three Neotropical *Lutra* are listed in Appendix 1 of C.I.T.E.S., the trade in these species being authorized only under exceptional circumstances. So much wild habitat remains in South America, yet stocks of otters are seriously depleted and this is due almost entirely to hunting and trapping for pelts.

The *Red Data Book* (Thornback and Jenkins, 1982) records that by 1967/1968, *L. felina* had been virtually exterminated in the region of Cape Horn and southern Tierra del Fuego, a region where Charles Darwin reported it as very plentiful in the 1830s. It is suggested, but without firm survey data, that the entire population of this species may number less than 1000. It is the most prized of all mammals by Chilean hunters, whereas in Peru it is persecuted mainly for the supposed damage it causes to freshwater prawns.

Lutra provocax is much restricted in range in Argentina, though it does exist in some remote and inaccessible areas, including national parks. Chéhebar (1985), carrying out a field survey in the Nahuel Huapi National Park, found signs at 28% of sites visited. Poaching was considered the main threat. In Chile, the status of *L. provocax* seems somewhat confused (Thornback and Jenkins, 1982), but the most recent assessment considers it to have been reduced to small, isolated and remote populations in south-central and southern Chile, where it is still heavily, and illegally, hunted (Miller *et al.*, 1983).

Lutra longicaudis has the widest distribution and is the least endangered of South American otters, with populations ranging through the uninhabited rain-forest of the interior. Nevertheless, the subspecies *platensis* is listed as endangered, while populations of other subspecies in Peru and Colombia are severely depleted (Brack-Egg, 1978; Donadio,

1978). Some 113 718 skins were exported from Peru during 1959–1972, with 14 544 in 1970 alone (Smith, 1981), while in Colombia, the annual legal kill in the early 1970s was between 6000 and 8000, but probably as many again were trapped illegally for their pelts. The killing of otters was banned in 1973 and official figures ceased, but trapping is thought to continue.

Trapping continues because of the high value of the pelt. In Argentina, a single pelt is worth almost three-quarters of the average monthly wage of a rural peasant, so that they will go to enormous lengths to procure an animal once it has been detected (Griva, 1978). In Chile, an otter skin is worth two or three months' wages to an unskilled worker, so that, although otters have been protected in that country since 1924, the economic incentive to hunt them is very high and the chance of being caught is very low (Miller *et al.*, 1983). The skins are transported to Argentina for export. Throughout South America there is a general lack of enforcement of wildlife legislation and little knowledge about wildlife or conservation in the populace at large. Without enforcement and education the future survival of South American otters, outside of strictly controlled reserves, looks bleak. A captive breeding programme is under way for the subspecies *platensis*.

Oriental otters

Four species of otters occur in the Oriental region, but none has been subjected to detailed studies in the field and what we know of their behaviour is based largely on observations of captive animals (e.g. Davis, 1968; Hodl-Rohn, 1974).

Lutra lutra is widely distributed through South-east Asia, with populations in southern India and Sri Lanka (Fig. 2.1). Wayre (1978) recorded the race *nair*, in Sri Lanka, from rice fields at sea level to mountain streams at altitudes of at least 1800 m. Freshwater crabs appeared to feature prominently in the diet. European populations of *L. lutra* are ecologically and behaviourally flexible (Chapter 2) and a study in the tropics would prove extremely interesting.

The hairy-nosed otter (*Lutra sumatrana*) resembles the European otter, but has its rhinarium covered with hair. Its range is shown in Fig. 6.8. Wayre (1974, 1978) considered that the hairy-nosed otter mainly inhabited mountain streams above 300 m, but Medway (1978) recorded it in the sea off Penang. Nothing is known of its ecology.

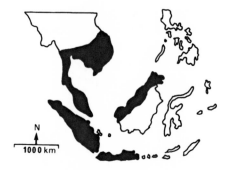

Fig. 6.8. Range of *Lutra sumatrana*.

The smooth-coated otter (*Lutra perspicillata*) has a markedly flattened tail and a short, smooth pelage (Fig. 6.9). It occurs from Pakistan, India and through South-east Asia, with an isolated race, *L. p. maxwelli*, in the marshes of southern Iraq (Fig. 6.10). It lives in estuaries, coastal mangrove swamps and large rivers, requiring undisturbed forest, mangrove or scrub adjacent to the water (Wayre, 1974). In Taman Negara National Park, a rainforest reserve in Malaysia, Pat Foster-Turley (pers. comm.) found smooth-coated otters common during a visit in 1984. She estimated a population of about 14 animals, including two groups of four to six, over a distance of 15 to 20 km of main river. Family groups are known to combine when hunting. Smooth-coated otters were also located in Malaysian mangrove swamps, from which individuals or pairs made ocasional forays into rice paddies.

According to Wayre (1978), smooth-coated otters feed mainly on fish of 15 to 30 cm length, with crabs being important in coastal areas, but no detailed studies of diet have been made. We have observed captive animals digging up and eating earthworms. In captivity, animals eat about 1 kg of food per day (Desai, 1974).

In Delhi Zoo, all matings of smooth-coated otters took place during August (Desai, 1974). The gestation period of four pregnancies was 61 to 62 days, with litter sizes ranging from two to five young. The females were seen to dig dens during their pregnancy and, after birth of the cubs, the father and other females were driven from the vicinity. The cubs were weaned at about 130 days, whereupon the male joined the family group and assisted in feeding the cubs with fish.

The smallest otter of the region is the Oriental small-clawed (or short-clawed) otter, *Aonyx cinerea* (Fig. 6.11*a,b*). The claws are rudimentary and peg-like. They feed like racoons, probing in mud and under stones (Davis, 1968).

Fig. 6.9. Smooth-coated otter, *Lutra perspicillata*.

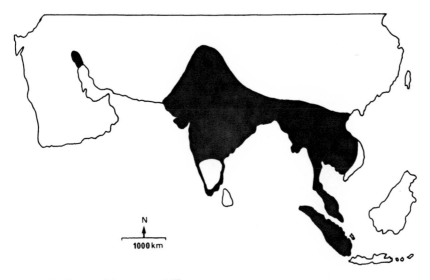

Fig. 6.10. Range of *Lutra perspicillata*.

Fig. 6.11.(*a*, *b*) Oriental small-clawed otter, *Aonyx cinerea*. (Photo (*a*): Dennis Hada.)

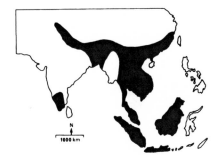

Fig. 6.12. Range of *Aonyx cinerea*.

Aonyx cinerea occurs in southern and north-east India, Bangladesh and South-east Asia (Fig. 6.12). They live in small streams, irrigation ditches and rice fields, in both upland and coastal areas. Pat Foster-Turley recorded groups of up to 15 in the paddies of Malaysia, where they were largely nocturnal, returning to dense bankside scrub during the day. The cycle of rice growing, with the paddies being periodically drained, is an important influence on the ecology of this highly social and very noisy otter. In coastal Sabah, Furuya (1976) observed them in groups of four to eight and they were active by day.

The small-clawed otter feeds mainly on small crabs, molluscs and small fish, such as gouramis and catfish (Wayre, 1978). Their habit of probing for food makes them unpopular with rice-growers because they uproot the growing plants (Pat Foster-Turley, pers. comm.), but, at the same time, they eat large numbers of small crabs, which are a pest in the rice fields (Khan, 1985).

The females come into oestrus every 28 days (Leslie, 1970). They do not, like smooth-coated otters, excavate their own dens, because of their weak claws, but use natural holes. Young are born after a gestation period of 60 to 64 days (Duplaix-Hall, 1975).

There is a long tradition in parts of India, China and South-east Asia of training otters to assist with fishing. All four species appear amenable to training and the practice, which probably originated in China, is described by Gudger (1927).

The otters discussed in this section do not appear to be in any immediate danger, in contrast to their relatives in other parts of the world. Much suitable habitat remains, while the agriculture is largely traditional. Nevertheless, deforestation is extensive in some areas (e.g. Malaysia) and opens up river valleys for exploitation, while increases in the use of persistent pesticides could prove highly damaging to populations. In some countries, such as Thailand, hunting for pelts has

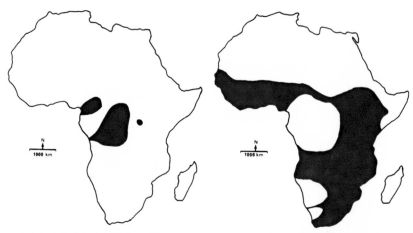

Fig. 6.13. Range of *Aonyx congica*.   **Fig. 6.14.** Range of *Aonyx capensis*.

Fig. 6.15. African clawless otter *Aonyx capensis*. Note the characteristic long vibrissae and the white cheeks, muzzle and throat. (Photo: David Rowe-Rowe.)

reduced populations. Much of the information on the distribution of Asian otters is based on old observations and modern field surveys are urgently required. Detailed field studies of these species would enable fascinating comparisons to be made with other otter species; without such studies any future conservation strategies will have to be based on an absolute dearth of ecological information.

Fig. 6.16. Spotted-necked otter *Lutra maculicollis*. Note short vibrissae and the dark cheeks and muzzle, compared with Fig. 6.15. (Photo: David Rowe-Rowe.)

African otters

Three species of otters are listed by Corbet and Hill (1980) as occurring in Africa south of the Sahara, the spotted-necked otter (*Lutra maculicollis*), the African clawless otter (*Aonyx capensis*) and the Zaïre clawless otter (*Aonyx congica*). Davis (1978) considers *A. congica* to be a subspecies of *A. capensis*. The clawless otter is similar in structure to the Asian small-clawed otter, but is much larger. It is born with tiny claws, but all are lost except for those on the middle three toes of the hind feet, which are used for grooming (Davis, 1968). The spotted-necked otter is a small beast and has a pale throat, neck and chest, irregularly spotted with dark markings.

Aonyx congica is distributed in the rivers draining the rainforests of central Africa (Fig. 6.13), but nothing is known of its ecology. *Aonyx capensis* (Fig. 6.15) has a much wider range and occurs in permanent streams and swamps (Fig. 6.14). Lakes and reservoirs appear to be a secondary habitat, at least in Natal (Rowe-Rowe, 1978*a*). In South Africa, the species also inhabits rocky sea coasts (Van der Zee, 1982) but is absent from the arid western interior. *Lutra maculicollis* (Fig. 6.16)

Fig. 6.17. Range of *Lutra maculicollis*.

Fig. 6.18. Typical habitat of both *Aonyx capensis* and *Lutra maculicollis* in Natal. It is an inland, grassland area, with good ground cover, no soil erosion and clear, unpolluted water. (Photo: David Rowe-Rowe.)

overlaps in both range (Fig. 6.17) and habitat (Fig. 6.18) with the clawless otter, but it prefers deeper water (Rowe-Rowe, 1978a) and is the commonest otter of the large lakes of central Africa. In South Africa, it is restricted to the eastern part of the country, being commonest in the upland regions, very scarce in the lowlands and absent from the sea coasts.

These African otters appear to have habits similar to those of other otter species, with distinctive sprainting places and rolling sites. The clawless otter is mainly crepuscular and partly noctural in its habits, whereas the spotted-necked otter is largely diurnal, with peaks of activity in early morning and early afternoon (Procter, 1963; Rowe-Rowe, 1978a). Both species regularly occur in groups of up to five animals, these larger gatherings presumably including both parents and the cubs. However, packs of up to 21 spotted-necked otters have been recorded on Lake Victoria by Procter (1963), who considered them to be made up of males making local movements.

The clawless otter is known to dig its own holt, whereas the spotted-necked otter does not (Rowe-Rowe, 1978a). On the South African coast, Van der Zee (1982) records 'holts' of clawless otter being situated adjacent to freshwater. From his description, these would be better described as 'couches', using British parlance, because the sleeping sites consisted of open areas under thick scrub. According to Tayler (1970) there are several shallow nests, interconnected by pathways under thick bushes. Above the nest foliage is pulled through the branches to form a basket-like roof. Each sleeping site was visited by an otter, on average once every three days, and the average distance between them was 1 km (Van der Zee, 1982). This population was estimated at one otter per 2 km of coastline, similar to that of Scottish coastal otters.

Little is known of the reproduction of African otters. Davis (1981) observed that captive spotted-necked otters always copulated in the water. Females have a gestation period of 58 to 63 days and one to two cubs are born, a litter size also reported in the wild in Natal (Rowe-Rowe, 1978a). Procter (1963) suggests that, in Tanzania, most cubs are born in September. In Natal, the litter size of the clawless otter is two to three (Rowe-Rowe, 1978a).

Quite detailed information on the diet and feeding behaviour of clawless and spotted-necked otters is now available, based on observation of both wild and captive animals. The clawless otter hunts both by sight and by feeling for prey beneath stones. Its long vibrissae may help it to find prey in murky water. Spotted-necked otters hunt by sight only. The dives of clawless otters were on average longer than those of spotted-necked otters feeding in the same lake (Rowe-Rowe, 1977a). Clawless otters capture most of their prey in their paws and frequently transport it in their forefeet, showing considerable manual dexterity (Fig. 6.19). Spotted-necked otters capture and transport prey in the mouth. Although clawless otters eat most of their food in the water, spotted-necked otters carry their prey onto land to eat (Rowe-Rowe,

Fig. 6.19. *Aonyx capensis* using its clawless fore-feet to hold prey, a crab. The head is tipped back in a characteristic way while chewing. (Photo: David Rowe-Rowe.)

1977*a*). Clawless otters have been observed using rocks, discarded bottles and a metal pipe as anvils to break open the shells of the freshwater mussel *Aspatharia wahlbergi*, when the molluscs were exposed in lake mud during a drought year (Donnelly and Grobler, 1976).

Rowe-Rowe (1977*b*) carried out food preference tests with captive clawless otters and found that they preferred crabs to frogs, and frogs to fish. When offered various fish species, they caught them in inverse proportion to the fishes' swimming ability (Rowe-Rowe, 1977*b*), as do other otter species. Small fish were captured more readily than large fish.

The diets of *L. maculicollis* and *A. capensis* have been studied in freshwater habitats in Natal (Rowe-Rowe, 1977*a*), and of the latter species in coastal habitat in Cape Province (Van der Zee, 1981) (Table 6.3). In the freshwater areas, spotted-necked otters took almost equal amounts of fish and crab, overall. Crabs, however, were the major item in spring, summer and autumn, with fish assuming importance in the winter. Crabs are inactive during the winter, remaining deeply hidden, while fish move more slowly in the cold weather. Casual observations from other parts of Africa indicate a much higher proportion of fish in the diet; the fish fauna of Natal is impoverished, so that local spotted-

Table 6.3. *Diet (relative frequency % of items in spraints) of otters in South Africa*

	Lutra maculicollis		*Aonyx capensis*		
	Trout water	Non-trout water	Trout water	Non-trout water	Sea coast
No. of samples	228	66	951	618	1129
Fish	38·2	25·9	2·8	8·3	52·5
Crustaceans	39·2	30·6	67·8	63·9	39·8
Octopus	—	—	—	—	7·5
Other invertebrates	1·8	5·6	2·0	4·2	0·3
Amphibians	20·2	27·8	25·2	21·2	0·2
Reptiles	—	—	0·4	1·1	—
Birds	0·6	10·2	1·4	1·3	—
Mammals	—	—	0·3	—	—

necked otters must have adapted to an increased diet of crabs. Frogs made up most of the remainder of the diet.

Crabs were much more important to *Aonyx*, which took relatively few fish in Natal. Frogs were also important and the diet confirms the results obtained from the preference experiments reported above. Frogs were taken in greater numbers in winter, when crabs became inactive.

The crabs taken by *Lutra* were generally small, whereas *Aonyx* took all sizes. Eels, although abundant in the rivers, were rarely taken, contrary to the situation with the European otter. The only other significant prey item was birds in the diet of spotted-necked otters in the non-trout water (10·2%). Elsewhere, *Aonyx* has been recorded catching flamingoes (*Phoeniconaias minor*) in Kenya (Rüppell and Rüppell, 1980) and has been accused of predating colonies of purple herons (*Ardea purpurea*) in Zimbabwe, but without conclusive evidence (Tomlinson, 1974). It has also been suggested that a commensal relationship may exist between *Aonyx* and the pied kingfisher (*Ceryle rudis*), the bird catching the fish that are disturbed by the foraging otter (Boshoff, 1978).

Rowe-Rowe (1977c) concluded that the food overlap between *Aonyx* and *Lutra* was 66%. Although introduced trout were taken when available, especially by *Lutra* (Table 6.3), the majority of fish were less than 20 cm, the legal minimum size for anglers to catch (Rowe-Rowe, 1978b), so that the impact of otters on these fisheries was negligible.

Mortimer (1963) observed that a captive spotted-necked otter, weighing 3·5 kg would eat about 650 g of food each day. Approximately 82% of this food was digested, with a daily spraint production of about 2·8 g dry weight.

Coastal *Aonyx* took a wide variety of prey species, but four species alone accounted over 80% of the diet (Table 6.3). These were two species of crabs (*Plagusia chabrus* and *Cyclograpsus punctatus*), an octopus (*Octopus granulatus*) and the suckerfish (*Chorisochismus dentex*). This latter sucks on to rocks, making it very vulnerable to otters. Other fish, though abundant, were taken in much smaller quantities, except under special circumstances, for instance when shoals became trapped in rock-pools.

In most African countries there is little information on the status and conservation requirements of otters. Rowe-Rowe (1985 and *in litt.*) contacted nature conservationists and zoologists throughout the range of otters. Information was returned for 25 countries (Table 6.4). *Aonyx capensis* was fairly common in nine countries and very rare in four, while

Table 6.4. *Status of otters in African countries*

| | Number of countries | |
Status	*Aonyx capensis*	*Lutra maculicollis*
Absent	3	7
Very rare	4	9
Rare	9	4
Fairly common	9	3
Common	0	3

Lutra maculicollis was fairly common, or common, in six countries, but very rare in nine. In Niger, for which Rowe-Rowe received no recent information, Poche (1973) reported *Aonyx capensis* as extinct and *L. maculicollis* as endangered.

Rowe-Rowe (1985) considers that, although otters are killed locally for food and skins, direct persecution is not a severe problem. Most African countries are members of C.I.T.E.S. and prevent the uncontrolled export of skins. The greatest problems are those associated with the increase in human population. In particular, overgrazing results in soil erosion and the siltation of rivers causes loss of the food of otters. The drainage of swamps to provide additional agricultural land, followed by increased pesticide usage, may also prove highly damaging to otter populations in the future.

The giant otter

Pteronura brasiliensis (Fig. 6.20) is one of the largest of South American carnivores, males attaining a weight of 26 to 32 kg and an overall length of 1·5 to 1·8 m, the females weighing 22 to 26 kg and measuring 1·5 to 1·7 m (Duplaix, 1980). Lengths of up to 2·4 m are recorded in earlier literature. The fur is generally chocolate brown in colour, with patterns of pale markings on the neck. The otter's head appears globular, due to the blunt, sloping muzzle, which sports long vibrissae. The long tail is dorso-ventrally flattened.

The giant otter was once widely distributed in South America (Fig. 6.21) but it has been much reduced due to poaching. It lives in large rivers and narrow forest creeks, mostly backed by dense vegetation (Fig. 6.22). Otters seem to prefer blackwater rivers, so called because the water

Fig. 6.20. Giant otter, *Pteronura brasiliensis*. (Photo: Nicole Duplaix.)

is stained with humic materials from decomposition on the forest floor. Such waters have little primary production, but the constant rain of animal and plant material from the surrounding forest supports dense fish populations. Duplaix (1980) considered that the major factors influencing giant otters' choice of habitat were low, sloping banks, with good cover and easy access to forest creeks or swampy areas, and an abundance of vulnerable prey in relatively shallow waters.

Giant otters are diurnal and social. In Surinam they were active between 06.30 and 18.30 hours, during which they were engaged in four main activities, which showed no particular periodicity (Duplaix, 1980). Figure 6.23 shows the proportion of time spent in these activities by an established pair of otters over five days. In Peru, Martha Brecht-Munn (1985) found that giant otters spent most of the morning feeding, during the middle of the day they rested and groomed, while the late afternoon was devoted to feeding and playing.

Resident giant otters live in family groups that are inquisitive and noisy, keeping up a constant hum of contact notes. Groups of otters usually consist of two adults, one or more sub-adults and one or more cubs, but, periodically, family groups may coalesce and travel together. In Surinam the largest group seen was of 16 animals, but 20 have been recorded together in Peru (Duplaix, 1980). Sub-adults probably stay with their parents until just before sexual maturity, when they leave to

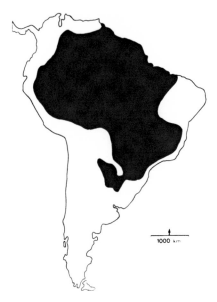

Fig. 6.21. Distribution of *Pteronura brasiliensis*. (From information in Melquist, 1984.)

Fig. 6.22. The giant otter, *Pteronura brasiliensis* on a camp site. (Photo: Nicole Duplaix.)

seek mates and form family units of their own. Most observations of single otters are probably these sub-adults, living a nomadic existence as transients within the population until they find a suitable partner.

Why should giant otter sub-adults stay with their parents? They could help their parents in rearing the next litter of offspring, they could

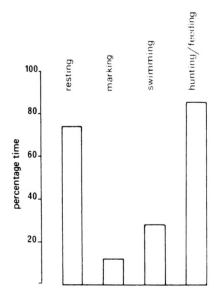

Fig. 6.23. Activity budget of a pair of giant otters over five days. (Adapted from data in Duplaix, 1980.)

co-operate in deterring predators or they could co-operate in capturing prey. Because of the relatedness of the individuals within a group, any assistance that enhances the survival of an individual will increase the genetic fitness of the helper (see Grafen, 1984, for a general discussion of inclusive fitness). Sub-adults do not provision cubs directly with food, as occurs, for example, with nest helpers of a number of bird species. They may, however, keep cubs out of danger by staying close to them and preventing them from straying (Duplaix, 1980). Sub-adults may, of course, learn from their parents some of the techniques in successful cub rearing, of value in their own future reproductive efforts. They do not, however, assist parents at the den (Duplaix, 1980).

Apart from man, giant otters have few predators, jaguars (*Panthera onca*), anacondas (*Eunectes murina*) and large caimans (*Caiman crocodylus*) being probably the most important. Lone otters appear especially vulnerable (Brecht-Munn, 1985) and unwary cubs may also be at risk. Social grouping may therefore be an anti-predator device, allowing both the earlier detection of a predator and concerted defence against it. Laidler and Laidler (1983) describe a group of six giant otters mobbing a large anaconda until it retreated into the forest.

The feeding success of groups of otters was higher than that of lone individuals (Laidler, 1982), though both Laidler (1982) and Duplaix (1980) considered that active co-operation in hunting did not occur, the animals merely benefiting from the general panic created in the fish shoal.

Nevertheless, Brecht-Munn (1985), in her Peruvian study area, observed that fishing success increased with increasing group size and that otters herded fish into shallow waters, observations also reported by Hershkovitz (1969). Clearly, more observations on fishing in groups are required.

Laidler (1982) observed two types of hunting: fishing in deep-water and in the shallows. Deep-water fishing involved vertical dives that usually lasted for less than one minute, but were occasionally up to two minutes long. Diving by the group was often fairly synchronized. When shallow-water fishing, otters excitedly and noisily chased after fish. Shallows fishing appeared more successful than deep-water diving.

Fish make up the greater proportion of the giant otter's diet. The rivers of South America are extremely rich in fish species and biomass. Duplaix (1980), in Surinam, made 202 direct observations of prey capture by two pairs of habituated otters and recorded that characoids (characins) made up 59·3%, siluroids (catfish) 22·6% and percoids (perch) 16·6% of fish caught, fish size ranging from 10 to 45 cm. Spraint analysis revealed that crabs were also frequently taken, while one male was seen twice to collect dung of a tapir (*Tapirus terrestris*) and eat it in the water. Characoids and catfish also made up the bulk of the diet of otters in Guyana (Laidler, 1982), while in Peru a mainly fish diet was supplemented by frogs, anacondas and small caimans (Brecht-Munn, 1985).

Pteronura shares its habitat with *Lutra longicaudis*, but *Lutra* is mainly crepuscular and nocturnal and appears to feed on smaller fish, so that competition is largely avoided.

Giant otters will deposit single spraints at sites similar to those used by other otter species, but such activity may be confined to transient individuals. Otter groups make use of camp sites, areas on the river bank cleared of vegetation by the animals. Duplaix (1980) found that camp sites could vary in length from 1·5 m to 28 m and in width from 1 m to 14·7 m and that they were rectangular or semi-circular in shape. They were positioned above high water mark and beneath overhanging trees, usually in a place that was visible from some distance. Within each camp site there is a latrine area, often about 1·5 m diameter, and several disused latrines, distinctive because of the thick layer of scales and bones, may also be present.

Both parents, but not the young, will mark the camp sites. The animals will break off twigs and leaves from bushes and trample these into the floor. Saplings are also trampled down and scent and urine are spread around the camp site as it is cleared. The otter defaecates on the

1977   1978

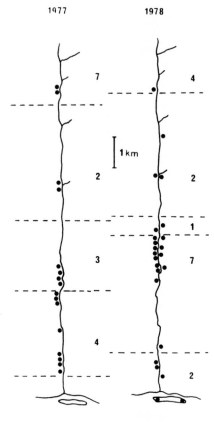

Fig. 6.24. Distribution of camp sites in use at Kaboeri Creek, Surinam, in March 1977 and March 1978, towards the end of the dry season. Estimates of group sizes along the creek are given, the horizontal dashed lines delineating apparent home-range boundaries. (After Duplaix, 1980.)

latrine area and the spraint is kneeded and trampled into the ground. The animals roll on the camp site and rub themselves with earth and may thus acquire a group scent. As well as marking, otters use the camp sites for resting and grooming.

Camp sites are fairly permanent features, almost half of those recorded by Duplaix (1980) being used in a subsequent year, and they develop a very strong, distinctive smell. In both Surinam (Fig. 6.24) and Guyana (Laidler, 1982), camp sites had a clustered distribution. They appear to serve a territorial function, advertising to wandering animals the presence of residents, using both olfactory and visual signals.

In her detailed study area at Kaboeri creek, Surinam, Duplaix (1980) considered that, during the dry season, otters limited their activities to well defined areas (Fig. 6.24). When reaching a given point in the creek, groups would always turn back. Individuals from the group occasionally travelled over the borders, but they would act very furtively. Otter groups marked their territories at camp sites along 2 to 3 km of creek. Groups would patrol part of the territory each day and they would call loudly if they heard another otter in the neighbourhood.

Social status is undoubtedly important in obtaining and holding a territory. In Kaboeri creek, one pair (the H group) lost their two cubs to a poacher. After this they were gradually displaced by a large family (the S group) of otters from downstream. The H female then disappeared and her mate retreated further to the top part of his territory. The S group eventually expanded into this area and the H male effectively became a transient.

During the wet season in Surinam, flooding of the surrounding forest takes place and the fish disperse and spawn within the floodwaters. It appears that the otters abandon their territories at this time and roam widely in the flooded forest.

In the Laidlers' (1983) study area in Guyana, seasonal flooding of creeks is limited due to the presence of a very large reservoir and dispersal of otters during the wet season does not occur. However, the apparent pattern of territory use is also different. Clusters of camp sites occur, as in Surinam, but are visited for periods of two to four days at intervals of two to three weeks (Laidler, 1982). The cluster of camp sites is considered a core area, used exclusively by a family group. It is suggested that, outside of the core area, otters travelled widely, different groups using the same water space, but probably at different times because aggressive encounters were never witnessed. Minimum home-ranges of 32 km of river or 20 sq. km of lake were estimated.

From the information presented, the land tenure systems in the Surinam and Guyana populations are clearly different. In both studies, otters could not be located for long periods, due to the very difficult working conditions. Further work, with radio–telemetry as a prerequisite, is required to assess the range of variation in territoriality in different populations.

In addition to camp sites, Duplaix (1980) also found dens in Surinam, although these were not observed in Guyana (Laidler, 1982). A den consisted of a tunnel, or tunnels, excavated into the river bank and leading to a chamber. One such den, with its entrances submerged at high tide, was occupied by four adults and three cubs. New dens were excavated by otters in the Kaboeri creek when they returned to their territories at the beginning of the dry season.

Females may use dens when giving birth to their cubs. Duplaix (1980) considered that most births in Surinam occurred at the beginning of the dry season, while Laidler (1982) thought there was no seasonality in Guyana, though data were scanty in both cases. The gestation period of giant otters is 65 to 70 days (Autuori and Deutsch, 1977) and litters generally contain one to three cubs, though five cubs were born to a captive female. From observations in the São Paulo zoo, Brazil, it was noted that the female took her offspring out of the den when they were 13 days old, they were placed in water at 20 days and they began to swim at 42 days. When 72 days old, they began to swim unaided and they ate fish when 90 days old.

Being diurnal and living in groups, giant otters are vulnerable to the depredations of man and this is exacerbated by the extremely inquisitive nature of the animal. At any unusual noise an otter will lift its head out of the water and begin periscoping, stretching and retracting its neck, trying to locate the source of the sound. Males may even charge across the water towards a boat, screaming loudly, making easy targets for a hunter.

Due mainly to hunting, the giant otter is classed as seriously endangered in Colombia, Brazil, Bolivia, Ecuador, Peru, Paraguay and Venezuela (Thornback and Jenkins, 1982). It is extinct, or on the verge of becoming so, in Uruguay and Argentina. Giant otters are known to be fairly widespread only in Surinam and Guyana.

Smith (1981) noted that a giant otter pelt was worth the equivalent of U.S.$50 to a peasant hunter in Brazil, as much as he could earn in ten days of gruelling forest labour, so the incentive to hunt is high. Between 1960 and 1969, 19 925 skins were legally exported from Brazil. From the smaller area of the Peruvian Amazon, 2017 pelts were exported in 1946, but during the 1960s the annual take was always less than 1000; over the period 1946 to 1973, 23 980 pelts were legally exported from Peru (Brack-Egg, 1978). In 1965, 1032 skins were exported from Colombia, but only 311 in 1970 (Donadio, 1978). As demand for skins remained constant or increased over these periods, overhunting was clearly occurring.

The giant otter was listed in Appendix 1 of C.I.T.E.S. in 1973, trade being strictly regulated and effectively banned for commercial purposes. Nevertheless, the extent of poaching and illegal trade is unknown. The rapid increase in human population in South America and the accelerating felling of tropical forest, itself resulting in erosion and associated water quality problems that damage fish stocks, point to a bleak future for the giant otter outside of strictly regulated reserves.

The sea otter

Sea otters (*Enhydra lutris*) (Fig. 6.25) were formerly widely distributed around the coasts of the northern Pacific Ocean, but excessive hunting in the eighteenth and nineteenth centuries reduced them to a few remnant populations. Protection during the present century has resulted in a marked increase in numbers and an expansion in range (Fig. 6.26).

Spending most of its time in the water, the sea otter has developed some seal-like characters. It is the heaviest of the Mustelidae, with males weighing from 27 to 38 kg and large animals reaching 45 kg. The females are much smaller, averaging 16 to 27 kg, with 32·6 kg recorded (Kenyon, 1981, 1982). However, they are very bulky, the total length being no more than 1·4 m, compared with 1·8 m of the lighter giant otter. The tail is very short. The forepaws are stumpy, but highly mobile and adapted for picking food from the sea bed and manipulating it while eating. The claws are retractile. The hind feet are large and flipper-like, used for propelling the otter through the water. The ears resemble those of seals rather than river otters. The sea otter has very large kidneys, which may be an adaptation to the salt-water habitat.

Unlike other marine mammals, the sea otter has no blubber layer to protect it from the cold sea environment. To compensate, the fur is much denser than that even of a fur seal (an adult male may have 800 million hairs!), and traps an insulating layer of air. To maintain the insulation, the otter spends much time cleaning and grooming. The otter also generates much internal heat to maintain body temperature, its basal metabolic rate being more than 2·5 times that of a mammal of similar size (Morrison, Rosemann and Estes, 1974).

The high basal metabolic rate requires a large intake of food, a sea otter consuming 20 to 30% of its body weight in a day or about 28 200 kJ per day (compared with, for example, 17 000 kJ per day for a manual worker). A fully grown male sea otter may consume 9·4 kg of food each day (Costa, 1978). Food passes through the otter in about three hours (Kenyon, 1969).

Sea otters feed mainly on or near the ocean floor, amongst rocks or on sandy or muddy substrates. Dives generally last from between 50 to 90 seconds (Limbaugh, 1961; Calkins, 1978) though Loughlin (1980a) records a dive of 265 seconds. Most foraging is carried out at a depth of around 20 m, though the maximum confirmed dive was to 97 m (Newby, 1975). The sea otter has a relatively large lung capacity as an adaptation to diving (Leith, 1976). With their dexterous hands, the

Fig. 6.25. The sea otter, *Enhydra lutris*, in a typical resting posture in the water, wrapped in kelp, Amchitka Island, Alaska. (Photo: James Estes.)

animals probe amongst rocks and seaweed, for prey, and they dig for invertebrates in soft sediments (Shimek, 1977). They have been observed pounding abalones with rocks to dislodge them from the sea floor (Houk and Geibel, 1974). Food is stored under the left arm to bring it to the surface.

Because the sea otter feeds largely on invertebrates, its teeth, uniquely among carnivores, have no sharp cutting edges. The rounded cusps are adapted to crush the hard exterior of its prey. Small prey items are

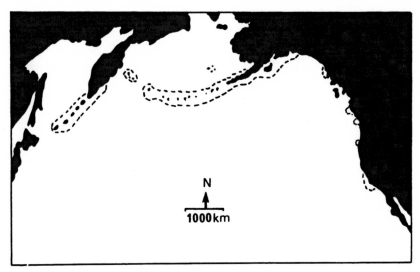

Fig. 6.26. Current range of *E. lutris*.

crushed and swallowed whole, but hard shells are broken with the aid of tools. The otter lies on its back in the water and bangs a shell against a rock, which it has collected from the sea bed, held on its chest. Prey items may also be knocked together. In Monterey Bay, California, McCleneghan and Ames (1976) watched a sea otter bringing discarded drinks cans to the surface. These were torn open to extract octopuses, which had made their homes within the cans.

Observations have suggested that foraging activity by sea otters is greatest during the early morning and late evening, both in California (Sandegren, Chu and Vandevere, 1973; Shimek and Monk, 1977) and in the northern Aleutian Archipelago (Kenyon, 1969; Estes and Smith, 1973). There is, however, much individual variation. Recent radio-telemetry studies in California (Loughlin, 1980*a*; Ribic, 1982*a*) have shown that otters are also active through the night. Activity is unrelated to time of day or tidal rhythm and there are generally three bouts of activity during a day. Californian otters were active for 46% of the time, 34% of which was devoted to feeding (Loughlin, 1980*a*).

Estes, Jameson and Rhode (1982) have recently made some fascinating comparisons of feeding activity and diet in three sea otter populations. Amchitka, in the western Aleutian archipelago of the north Pacific, holds a large sea otter population, which has probably been at equilibrium density for about 30 years. Attu, also in the Aleutian archipelago, had no otters until the early 1960s and Blanco Reef, Oregon, had no otters

Table 6.5. *Percentage of daytime spent in various activities by sea otters in three study areas (after Estes et al., 1982)*

Locality	No. obser-vations	Foraging	Resting	Grooming	Other
Attu, Aleutians	6669	16·3	53·5	14·9	15·2
Blanco Reef, Oregon	3762	16·8	59·3	16·7	7·2
Amchitka Island, Aleutians	8385	54·0	32·7	5·9	7·4

until 1970, when animals were translocated from Amchitka. Table 6.5 shows that sea otters spent much longer foraging at Amchitka than at the other two sites, with correspondingly less time in other activities. Furthermore, the Amchitka otters showed a well-defined bi-modal pattern of foraging activity, feeding most just after dawn and again in the late afternoon. The Attu and Oregon otters lacked any pattern in foraging activities.

There were also differences in the diet. While sea otters at all localities ate mainly sea urchins, their preferred food, the Amchitka otters also took many fish (12·3%, compared with 0·2% at Attu and none in Oregon). Dives for food were successful twice as often at Attu than at Amchitka.

Estes *et al.* (1982) suggested that, at Attu and Blanco Reef, sea otters were well below their equilibrium density and their preferred prey, sea urchins, were abundant and could be caught easily at any time of day. At Amchitka, the equilibrium population of otters has reduced the numbers of sea urchins, which has encouraged the growth of kelp beds (see below). This has resulted in an increase in fish which, although serving as an alternative prey for otters, are more elusive than sea urchins. Fish are most vulnerable at twilight, thus explaining the crepuscular feeding activity of Amchitka otters.

Table 6.6 presents data from selected studies on the diet of otters. Sea otters have been described as opportunistic feeders of invertebrates, but they appear to have preferences for those species with a higher energy content (Costa, 1978). Thus, in unexploited habitats in California they feed mainly on sea urchins and abalones. Sea urchins largely disappear from the diet after the otters have been present in an area for about a year and there is an increased use of kelp crabs and the appearance of clams (Ostfield, 1982). The diet is eventually expanded to include a

Table 6.6. *Diet of sea otters*

Locality	Method	Sample size	Percentage occurrence of:					Reference
			Sea urchins	Molluscs	Crabs/ lobsters	Fish	Others	
U.S.S.R.	Faeces	1480	59·0	23·0	10·0	6·7	1·0	Barabash–Nikiforov, 1947
Amchitka Island	Stomach	309	21·0	31·0	0·4	22·0	22·0	Kenyon, 1969
Alaska	Direct observation	597	0	81·9	7·0	0	10·9	Calkins, 1978
California	Direct observation	243	0	69·2	25·9	0	4·9	Ebert, 1968
California	Direct observation	455	32·8	51·1	15·1	0·2	0·8	Hall and Schaller, 1964
California	Direct observation	748	14·6	28·5	53·8	0	3·1	Ostfield, 1982

variety of invertebrates (Wild and Ames, 1974). As described above they will also take fish, while birds are also occasionally killed and eaten (Van Wagenen, Foster and Burns, 1981). Sea otters prefer food species that yield a high energy intake per unit of foraging time and, as such items become depleted, they switch to the species with the next highest yield (Ostfield, 1982).

The predation of sea otters on herbivores has a marked effect on the structure of the shallow water communities in the North Pacific. Where sea otters are scarce, sea urchins, chitons and other grazers are abundant and large, and the bare, rocky substrate is carpeted with bivalves, tube worms and their predators. By islands where sea otters are abundant, sea urchins are small and there is a luxuriant growth of kelp. Sessile invertebrates, such as mussels and barnacles, are scarce, but the detritus derived from the kelp supports numerous amphipods and prawns, which themselves form the food supply of a large population of nearshore fishes, finding refuge and breeding sites within the seaweeds (Estes, Smith and Palmisano, 1978). Islands without otters also have fewer vertebrates such as seals (*Phoca vitulina*) and bald eagles (*Haliaeetus leucocephalus*), which also rely directly or indirectly on fish (Estes and Palmisano, 1974). The effects of sea urchin grazing on kelp communities have been verified by experimental manipulation of urchin populations (Duggins, 1980; Palmisano, 1983). The relationships between sea otters, nearshore communities and prehistoric Aleut aborigines have also been examined (Simenstad, Estes and Kenyon, 1978). An analysis of stratified Aleut middens, representing prey remains of 2500 years, has shown dramatic changes in diet. Simenstad *et al.* have argued that these changes were the result of local overexploitation of sea otters. As sea otters were exterminated, Aleut diet switched to one dominated by invertebrates rather than fish. The sea otter is clearly a keystone predator in the inshore waters of the North Pacific.

Apart from man, the sea otter has few enemies. Young sea otters may be eaten by bald eagles. Sherrod, Estes and White (1975) recorded incidents of otter cubs being plucked from the sea by eagles in the Aleutian Islands. An examination of 114 nests over a five-year period revealed that sea otter remains made up 15·6% of the total remains, though many of these must have been of dead cubs scavenged from the shore. In California, sea otters may be killed by white sharks (*Carcharodon carcharias*). Ames and Morejohn (1980) attribute 9 to 15% of the 657 sea otter mortalities reported to the California Department of Fish and Game to sharks.

Advancing pack-ice may cause mortality in northern populations of sea otters. Schneider and Faro (1975) reported that, in 1971, sea ice advanced rapidly further south than normal and sea otters were trapped as ice formed around them. They were forced to travel across ice and land in search of open water, and many died of malnutrition. In the following year, the ice advanced more slowly and the otters were able to move south on the edge of the ice, so that, although the distribution of the population was temporarily altered, there were few deaths.

Unlike other mustelids, scent marking can play no role in the social organization of sea otters, because they spend so much time in the water, and this species has no scent glands. Because they are active during the day, direct observation of social interactions can be made and these have been facilitated by tagging sea otters on their hind feet and ears (Ames, Hardy and Wendell, 1983). Animals have also been followed by radio-telemetry.

Populations of sea otters show distinct sexual segregation, rafts of males and females occurring in different locations, the female aggregations being generally larger and occupying more protected areas with better food supplies (Estes, 1980). However, some males establish territories, defended areas, associated with female rafts.

In Alaska, two males had territories of, respectively, 1·25 sq. km and 0·75 sq. km (Calkins and Lent, 1975). Loughlin (1980b), in California, recorded male territories averaging 0·35 sq. km (range 0·18–9·58 sq. km, $n = 4$). Non-territorial male home-ranges were larger, averaging 0·44 sq. km (range 0·29–1·38 sq. km, $n = 7$), with female home-ranges being larger still, averaging 0·8 sq. km (range 0·28–1·98 sq. km, $n = 8$). Home-ranges were rectangular, averaging 2·5 km long, parallel with the shore and extending seawards for about 0·3 km. Ribic (1982b), at another Californian site, found home-range lengths longer than those reported by Loughlin (1980b). This resulted in larger home-range areas, those of males averaging 0·8 sq. km and of females 2·7 sq. km. Home-range size may be a function of resources.

Sea otters have generally been considered as non-migratory, but at least some members of the population may make seasonal movements and wander long distances. Odemar and Wilson (1969) observed that 5 of 17 translocated otters moved 72 km back to their point of capture in two months, while F. Wendell (in Ribic, 1982b) recorded one sea otter moving 48 km in less than 22 hours. Ribic (1982b) observed three male otters moving 80 km south along the California coast to the southern edge of the population's range, one animal travelling 70 km in two days. It

Fig. 6.27. Female *E. lutris*, with her single, recently born cub, hauled out on the upper intertidal zone at Amchitka Island, Alaska. (Photo: James Estes.)

was suggested that male sea otters may be seasonally migratory, but females are resident.

The sea otter is polygynous and mating may occur without prior pair-bonding taking place (Vandevere, 1970). In some populations there is delayed implantation (Kenyon, 1982), with a total gestation period of about eight months. However, Vandevere (1983) recorded a gestation period of four months for the California sea otter and suggested that delayed implantation does not occur in this population. At high population densities, females may breed only every second year, but they breed annually when populations are low (Loughlin, Ames and Vandevere, 1981; Vandevere, 1983).

Breeding and pupping may occur throughout the year, but in Alaska there are peaks of births in May and June (Schneider, 1972), while most

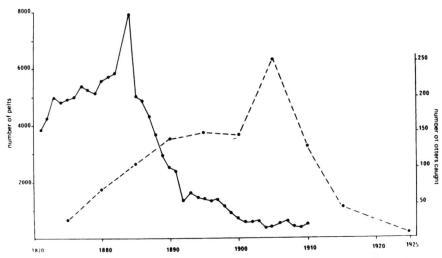

Fig. 6.28. Pelts (solid line) of *E. lutris* placed on the London Auction Market 1871–1910, and *E. lutris* caught (broken line) on the Commander Islands, 1870–1924, expressed as five-year moving averages. (Adapted from Barabash-Nikiforov, 1947.)

young in California are born between December and February (Vandevere, 1970; Sandegren *et al.*, 1973). Most of the young are probably born in the water, though terrestrial births have been reported (Kenyon, 1982). A single cub is born (Fig. 6.27). Females with cubs are aggressive, especially during bad weather. The young begin to take solid food when about 17 days old (Vandevere, 1983) and they start diving when about two months old (Sandegren *et al.*, 1973). Cubs are dependent on the mother for about eight months. Vandevere (1983) found that a female attracted a male, who assisted in the weaning of the cub. Females apparently reach sexual maturity at about four years (Kenyon, 1969), and males at about five or six years, though they do not become active breeders until several years later.

The sea otter was originally widely distributed around North Pacific coasts, from Baja California northwards to the Alaska peninsula, across the Aleutian, Pribilof and Commander islands to the Kamchatka peninsula, U.S.S.R., and southwards to Hokkaido, Japan (Estes, 1980). However, the species was extensively hunted for its fur in the eighteenth and nineteenth centuries and by 1911 the distribution was reduced to 13 known remnant populations, when it was given protection under the International Fur Seal Treaty. The extent of the destruction of sea otters can be inferred from Fig. 6.28, which shows the number of pelts sold on the London market between 1871 and 1910; a very sharp decline

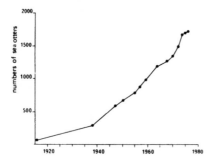

Fig. 6.29. Growth of the population of *E. lutris* in California since 1914, from estimates carried out by the Californian Department of Fish and Game.

occurred after 1884. Also shown are the official catch records of the Commander Islands, which reached a peak in the early 1900s and then went into sharp decline, the species finally receiving protection here in 1924.

Following protection, the sea otter population responded well. During the early stages of recovery, the annual rate of increase in Alaska was estimated at 19%, subsequently decreasing to 5% as populations approached equilibrium (Lensink, 1960). In 1972, the Alaskan population of sea otters was estimated at 100000 to 120000, with about half of the range re-occupied (Schneider, 1978). The population in the early 1980s is about 150000 to 200000 animals, assuming an overall rate of increase of 5% (Johnson, 1982). The California sea otter population has also increased (Fig. 6.29) to about 1800 individuals (Davis, 1977).

Population increases and range expansion in sea otters have been assisted by translocations from Alaska. Initial attempts were unsuccessful, probably because the fur became soiled and matted during transit and lost its insulatory qualities. However, between 1965 and 1972, 708 sea otters were translocated to other sites in Alaska, British Columbia, Washington and Oregon (Jameson *et al.*, 1982). Those translocations to South-east Alaska were the most successful. The British Columbia population is breeding, but has not increased, remaining at about 70 animals (Morris, Ellis and Emerson, 1981). The Washington transplant is also breeding, but remains small, while those to the Pribilof Islands and the Oregon coast appear to have failed. It seems that the number of sea otters at a transplant site decreases dramatically soon after release, due largely to emigration. When the translocation consists of few animals, those that remain after emigration and mortality may be too few for reproduction to outweigh losses due to emigration and mortality. Jameson *et al.* (1982) have suggested that a large nucleus population is required for successful establishment and recommended 25 to 30 animals per year over three to five years. In considering possible future translocations of Californian sea otters, Ralls *et al.* (1983) recommended a similar translocation strategy as the minimum necessary to retain most of the genetic diversity in the founder population.

The recovery of the sea otter populations can be considered as a success story, but the California population in particular is still small and vulnerable, especially from oil pollution (Van Blaricom and Jameson, 1982). Increases in population size have also led to predictable conflicts with fishermen. In California, the sea otter has been blamed for substantial declines in abalones and clams and it has been suggested that sea otters should be excluded from areas set aside as shellfisheries. Others argue that the decline in shellfish is largely due to human exploitation of luxury foods. The Friends of the Sea Otter, established in 1968, consider that otter populations should remain fully protected and be allowed to spread naturally, without restrictions in range and assisted by translocations (Davis, 1979).

In Alaska, a similar conflict between sea otters and fisheries is developing (Johnson, 1982), where both sea otter populations and commercial harvesting of shellfish have been increasing. Experimental harvests of about 2000 sea otters were made between 1967 and 1972, when they became protected under the Sea Mammal Protection Act. An attempt to sell skins on the Seattle Fur Exchange apparently yielded a poor return. Kenyon (1982) suggests that the Alaskan population would support an annual harvest of several thousand skins. Future conflicts between those viewing sea otters in different lights seem likely.

References

Abouzaid, H. and Hajji, A. (1982). L'eau potable au Maroc. *Aqua*, 2, 13–18.

Adams, W. M. (1984). *Implementing the Act. A study of habitat protection under Section II of the Wildlife and Countryside Act 1981*. British Association of Nature Conservationists/World Wildlife Fund, Oxford.

Ames, J. A., Hardy, R. A. and Wendell, F. E. (1983). Tagging materials and methods for sea otters, *Enhydra lutris*. *Calif. Fish Game*, 69, 243–252.

Ames, J. A. and Morejohn, G. V. (1980). Evidence of white shark, *Carcharodon carcharias*, attacks on sea otters, *Enhydra lutris*. *Calif. Fish Game*, 66, 196–209.

Anděra, M. and Trpák, P. (1981), Škodná nebo predátor? *Památky a Příróda*, 10, 609–618.

Anderson-Bledsoe, K. L. and Scanlon, P. E. (1983). Heavy metal concentrations in tissues of Virginia river otters. *Bull. Environ. Contam. Toxicol.*, 30, 442–447.

Andrews, E. and Crawford, A. K. (1984). *Wye tributaries otter survey*. Vincent Wildlife Trust, London.

Anglian Water Authority (1984). Work starts on ORSU's. *Anglian Fisheries News*, no. 7, 8.

Anonymous (1969). The otter in Britain. *Oryx*, 10, 16–22.

Anonymous (1974). The otter in Britain – a second report. *Oryx*, 12, 429–435.

Anonymous (1979). *Wildlife introductions to Great Britain*. Nature Conservancy Council, London.

Aulerich, R. J. and Ringer, R. K. (1970). Some effects of chlorinated hydrocarbon pesticides on mink. *Am. Fur Breeder*, 43, 10–11.

Aulerich, R. J. and Ringer, R. K. (1977). Current status of PCB toxicity to mink, and effect on their reproduction. *Arch. Environ. Contam. Toxicol.*, 6, 279–292.

Autuori, M. P. and Deutsch, L. A. (1977). Contribution to the knowledge of the giant Brazilian otter, *Pteronura brasiliensis* (Gmelin, 1788), Carnivora, Mustelidae: rearing in captivity. *Zool. Garten*, 47, 1–8.

Baker, J. R., Jones, A. M., Jones, T. P. and Watson, H. C. (1981). Otter (*Lutra lutra*) mortality and marine oil pollution. *Biol. Conserv.*, **20**, 311–321.

Barabash-Nikiforov, I. I. (1947). *Kalan (The sea otter)*. Soviet Ministrov RSFSR. [translated by Israel Program for Scientific Translation, Jerusalem].

Baruš, Š. and Zejda, J. (1981). The European otter (*Lutra lutra*) in the Czech Socialist Republic. *Acta Sc. Nat. Brno*, **15**, 1–41.

Bas, N., Jenkins, D. and Rothery, P. (1984). Ecology of otters in northern Scotland. V. The distribution of otter (*Lutra lutra*) faeces in relation to bankside vegetation on the River Dee in summer 1981. *J. Appl. Ecol.*, **21**, 507–513.

Baumgart, G. (1977). *Densité et distribution de quelques carnivores d'Alsace en zone forestiere*. Laboratoire d'Ecologie, Université Louis Pasteur, Strasbourg.

Berry, R. J. (1983). Genetics and conservation. In Warren, A. and Goldsmith, F. B. (eds.), *Conservation in perspective*, pp. 141–156. John Wiley and Sons, Chichester.

Birks, J. D. S. and Linn, I. J. (1982). Studies of home range of the feral mink, *Mustela vison*. *Symp. Zool. Soc. Lond.*, **49**, 231–257.

Blackmore, D. K. (1963). The toxicity of some chlorinated hydrocarbon insecticides to British wild foxes (*Vulpes vulpes*). *J. Comp. Pathol. Ther.*, **73**, 391–409.

Blas-Aritio, L. (1978). Informe sobre la situación de la nutria en España. In Duplaix, N. (1978), *loc. cit.*, pp. 140–142.

Boitani, L. (1976). *Reintroductions: techniques and ethics*. World Wildlife Fund, Rome.

Boshoff, A. F. (1978). Possible pied kingfisher–clawless otter commensalism. *Ostrich*, **4**, 89.

Bouchardy, C. (1981). *La loutre (Lutra lutra) L.* Centre Ornithologique Auvergne, Societé pour l'etude et la protection de la faune sauvage et des milieux naturels, Clermont-Ferrand.

Bourne, W. R. P. (1978). Mink and wildlife. *BTO News*, **91**, 1–2.

Brack-Egg, A. (1978). Situación actual de las nutrias en el Peru. In Duplaix, N. (1978), *loc. cit*, pp. 76–84.

Bradshaw, A. D. (1982). Presidential viewpoint. *Bull. Brit. Ecol. Soc.*, **13**, 112–114.

Braun, A.-J. (1984). A brief history of *Lutra lutra* L. in Brittany and present status and distribution. Unpublished report to Groupe Loutres, France.

Brecht-Munn, M. (1985). The giant otter in Peru. In Duplaix, N. and Kempf, C. (1985), *loc. cit.*, in press.

Brooker, M. P. (ed.). (1982). *Conservation of wildlife in river corridors*. Part I. *Methods of survey and classification*. Welsh Water Authority, Brecon.

Brooker, M. P. (ed.). (1983). *Conservation of wildlife in river corridors*. Part II. *Scientific assessment*. Welsh Water Authority, Brecon.

Brooker, M. P., (1984). Conservation of wildlife in river corridors. *Nature in Wales*, **2**, 11–20.

Brookes, A., Gregory, K. J. and Dawson, F. H. (1983). An assessment of river channelization in England and Wales. *Sci. Total Environ.*, **27**, 97–111.

Brown, B. T., Warren, P. L., Anderson, L. S. and Gori, D. F. (1982). A record of the southern river otter, *Lutra longicaudis*, from the Rio Yaqui, Sonora, Mexico. *J. Ariz. Nev. Acad. Sci.*, **17**, 27–28.

Broyer, J., Aulagnier, S., Destre, R. and Gaschignard, O. (1984). La loutre *Lutra lutra splendida* dans le sud du Maroc. *Bull. Groupe Loutres*, **18**, 12–20.

Bunyan, P. J. and Stanley, P. I. (1982). Toxic mechanisms in wildlife. *Reg. Toxicol. Pharmacol.*, **2**, 106–145.

Burton, R. (1979). *Carnivores of Europe*. Batsford, London.

Cabello, C. C. (1978). La nutria de mar (*Lutra felina*) en la Isla de Chiloe, Chile. In Duplaix, N. (1978), *loc. cit.*, pp. 108–118.

Cabello, C. C. (1983). La nutria de mar en la Isla de Chiloe. *Corporacion National Forestal, Bol. Tec.* **6**, 1–37.

Cabrera, A. (1932). Los Mamíferos de Marruecos. *Trabajos del Museo Nacional de Ciencias Naturales, Madrid, Ser. Zoológica*, **57**, 145–147.

Cagnolaro, L., Rosso, D., Spagnesi, M. and Venturi, B. (1975). Inchiesta sulla distribuzione della Lontra (*Lutra lutra* L.) in Italia e nei Cantoni Ticino e Grigioni (Svizzera) 1971–1973. *Laborat. Zool. appl. Caccia. Biolog. Selvaggina*, **63**, 1–120.

Calkins, D. G. (1978). Feeding behaviour and major prey species of the sea otter *Enhydra lutris*, in Montague Strait, Prince William Sound, Alaska. *Fishery Bull.*, **76**, 125–131.

Calkins, D. and Lent, P. C. (1975). Territoriality and mating behaviour in Prince William Sound sea otters. *J. Mammal.*, **56**, 528–529.

Callejo Rey, A. and Delibes, M. (1985). L'alimentation de la loutre (*Lutra lutra*) en Espangne. In Duplaix, N. and Kempf, C. (1985), *loc. cit.*, in press.

Callejo Rey, A., Guitan Rivera, J., Bas Lopez, S., Sanchez Canals, J. L. and Castro Lorenzo, A. D. E. (1979). Primeros datos sobre la dieta de la nutria, *Lutra lutra* (L.), en aguas continentales de Galicia. *Doñana, Acta Vertebr.*, **6**, 191–202.

Camby, A., Le Gall, O. and Maizeret, C. (1984). Atlas d'identification des restes alimentaires de la loutre (premiers elements). *Bull. Groupe Loutres, France*, **16**, 1–31.

Cassola, F. (1979). Shooting in Italy: the present situation and future perspectives. *Biol. Conserv.*, **16**, 85–106.

Cassola, F. (1980). The status and legal position of the otter *Lutra lutra* L. in Italy. *Otters, J. Otter Trust*, 1980, pp. 23–25.

Chadwick, M. J. (1983). Acid depositions and the environment. *Ambio*, 12, 80–82.

Chanin, P. (1981). The diet of the otter and its relations with the feral mink in two areas of south-west England. *Acta Theriol.*, 26, 83–95.

Chanin, P. R. F. and Jefferies, D. J. (1978). The decline of the otter *Lutra lutra* L. in Britain: an analysis of hunting records and discussion of causes. *Biol. J. Linn. Soc.*, 10, 305–328.

Chapman, P. J. and Chapman, L. L. (1982). *Otter survey of Ireland 1980–81*. Vincent Wildlife Trust, London.

Chéhebar, C. E. (1985). A survey of the southern river otter *Lutra provocax* Thomas in Nahuel Huapi National Park, Argentina. *Biol. Conserv.*, 32, 299–307.

Chesser, R. K., Smith, M. H. and Brisbin, I. L. (1980). Management and maintenance of genetic variability in endangered species. *Int. Zoo Yb.*, 20, 146–154.

Chitampalli, M. B. (1979). Miscellaneous notes. 1. On the occurrence of the common otter in Maharashtra (Itiadoh Lake–Bhandara District) with some notes on its habits. *J. Bombay Nat. Hist. Soc.*, 76, 151–152.

Clark, D. R. (1981). Bats and environmental contaminants: a review. *United States Department of the Interior, Fish and Wildlife Service, Special Scientific Report – Wildlife*, No. 235, Washington D.C.

Clark, D. R., Laval, R. K. and Tuttle, M. D. (1982). Estimating pesticide burdens of bats from guano analyses. *Bull. Environ. Contam. Toxicol.*, 29, 214–220.

Clark, R. B. (1984). Impact of oil pollution on seabirds. *Environ. Pollut. Ser. A.*, 33, 1–22.

Clayton, C. J. and Jackson, M. J. (1980). Norfolk otter survey 1980–1981. *Otters, J. Otter Trust*, 1980, pp. 16–22.

Cocks, A. H. (1881). Note on the breeding of the otter. *Proc. Zool. Soc. Lond.*, 1881, pp. 249–250.

Cooke, B. K. and Stringer, A. (1982). Distribution and breakdown of DDT in orchard soil. *Pestic. Sci.*, 13, 545–551.

Corbet, G. B. and Hill, J. E. (1980). *A world list of mammalian species*. British Museum and Cornell University Press.

Corbet, G. B. and Southern, H. N. (1977). *The handbook of British mammals*. Blackwell, Oxford.

Costa, D. (1978). The sea otter: its interaction with man. *Oceanus*, 21, 24–30.

Crawford, A., Jones, A. and McNulty, J. (1979). *Otter survey of Wales 1977–78*. Society for the Promotion of Nature Conservation/Nature Conservancy Council, London.

Criel, D. (1984). De situatie van de otter *Lutra lutra* in Belgie: een vooronderzoek. *Lutra*, **27**, 36–41.

Crisp, D. T., Mann, R. H. K. and McCormack, J. C. (1974). The population of fish at Cow Green, Upper Teesdale, before impoundment. *J. Appl. Ecol.*, **11**, 969–996.

Crisp, D. T., Mann, R. H. K. and McCormack, J. C. (1975). The population of fish in the River Tees system on the Moor House National Nature Reserve, Westmorland. *J. Fish Biol.*, **7**, 573–593.

Crofton, K. M., Taylor, P. H., Bull, R. J., Sivulka, D. J. and Lutkenhoff, S. D. (1980). Developmental delays in exploration and locomotor activity in male rats exposed to low level lead. *Life Sci.* **26**, 823–831.

Cumbie, P. M. (1975). Mercury levels in Georgia otter, mink and freshwater fish. *Bull. Environ. Contam. Toxicol.*, **14**, 193–197.

Cummins, K. W. (1979). The natural stream ecosystem. In Ward, J. V. and Stanford, J. A. (eds.), *The ecology of regulated streams*, pp. 7–24. Wiley, New York.

Cutler, J. R. (1981). *Review of pesticide usage in agriculture, horticulture and animal husbandry 1975–79. Survey Report* no. 27. Department of Agriculture and Fisheries for Scotland, Edinburgh.

Danilov, P. I. and Tumanov, I. L. (1975). The reproductive cycles of some Mustelidae species. *Byull. Mosk. O-va Ispyt. Oto. Biol.*, **80**, 35–47.

Davis, B. S. (1977). The southern sea-otter revisited. *Pacific Discovery*, **30**, 1–13.

Davis, B. S. (1979). Friends of the Sea Otter; position paper on management perspectives. *Otter Raft*, **22**, 3.

Davis, J. A. (1968). River otters, sea otters and part-time otters, part 2. *Anim. Kingdom*, **71**, 8–12.

Davis, J. A. (1978). A classification of otters. In Duplaix, N. (1978), *loc. cit.*, pp. 14–33.

Davis, J. A. (1981). Breeding the spot-necked otter. Abstract, Second International Otter Colloquium, Norwich, September 1981.

Dawson, F. H. and Haslam, S. M. (1983). The management of river vegetation with particular reference to shading effects of marginal vegetation. *Landscape Planning*, **10**, 147–169.

Day, M. G. (1966). Identification of hair and feather remains in the gut and faeces of stoats and weasels. *J. Zool., Lond.*, **148**, 201–217.

Delibes, M. and Callejo, A. (1985). On the status of the otter in Spain. In Duplaix, N. and Kempf, C. (1985), *loc. cit.*, in press.

Desai, J. H. (1974). Observations on the breeding habits of the Indian smooth otter *Lutrogale perspicillata* in captivity. *Int. Zoo Yb.*, **14**, 123–124.

Descy, J.-P. and Empain, A. (1984). Meuse. In Whitton, B. A. (ed.), *Ecology of European rivers*, pp. 1–23. Blackwell Scientific Publications, Oxford.

Dollinger, P. (1983). Legislation. *Naturopa*, **45**, 10–12.

Donadio, A. (1978). Otter trade and legislation in Colombia. In Duplaix, N. (1978), *loc. cit.*, pp. 34–42.

Donnelly, B. G. and Grobler, J. H. (1976). Notes on food and anvil using behaviour by the Cape clawless otter, *Aonyx capensis* in the Rhodes Matopos National Park, Rhodesia. *Arnoldia Rhod.*, 7, 1–8.

Duggins, D. O. (1980). Kelp beds and sea otters: an experimental approach. *Ecology*, 61, 447–453.

Duplaix, N. (ed.) (1978). *Otters: proceedings of the first working meeting of the Otter Specialist Group, Paramaribo, Surinam, March 1977.* International Union for the Conservation of Nature and Natural Resources, Morges, Switzerland.

Duplaix, N. (1980). Observations on the ecology and behavior of the giant river otter *Pteronura brasiliensis* in Surinam. *Rev. Ecol. (Terre Vie)*, 34, 495–620.

Duplaix, N. and Kempf, C. (eds.) (1985). Proceedings of the 3rd International Otter Colloquium, Strasbourg, November 1983, in press.

Duplaix-Hall, N. (1975). River otters in captivity: a review. In Martin, R. D. (ed.), *Breeding endangered species in captivity*, pp. 315–327. Academic Press, London.

Ebert, E. E. (1968). A food habits study of the southern sea otter, *Enhydra lutris nereis*. *Calif. Fish Game*, 54, 33–42.

Eckstein, M. I. (1984). *Resource evaluation and conservation studies within river corridors.* Occasional Report OR16. Water Research Centre, Marlow, Bucks.

Edwards, C. A. (1966). Insecticide residues in soil. *Resid. Rev.*, 13, 83–132.

Elliot, K. (1983). The otter (*Lutra lutra* L.) in Spain. *Mammal Rev.*, 13, 25–34.

Elmhirst, R. (1938). Food of the otter in the marine littoral zone. *Scot. Nat.*, 1938, pp. 99–102.

Erlinge, S. (1967a). Food habits of the fish-otter *Lutra lutra* L. in South Swedish habitats. *Viltrevy*, 4, 371–443.

Erlinge, S. (1967b). Home range of the otter *Lutra lutra* L. in southern Sweden. *Oikos*, 18, 186–209.

Erlinge, S. (1968a). Food studies on captive otters (*Lutra lutra* L.). *Oikos*, 19, 259–270.

Erlinge, S. (1968b). Territoriality of the otter *Lutra lutra* L. *Oikos*, 19, 81–98.

Erlinge, S. (1969). Food habits of the otter *Lutra lutra* L. and mink *Mustela vison* Schreber in a trout water in southern Sweden. *Oikos*, 20, 1–7.

Erlinge, S. (1972a). The situation of the otter population in Sweden. *Viltrevy*, 8, 379–397.

Erlinge, S. (1972b). Interspecific relations between otter *Lutra lutra* and mink *Mustela vison* in Sweden. *Oikos*, 23, 327–335.

Erlinge, S. (1979). Adaptive significance of sexual dimorphism in weasels. *Oikos*, 33, 233–245.

Erlinge, S. (1980). Die Situation des Fischotters in Schweden. In Reuther, C. and Festetics, A. (1980), *loc. cit.*, pp. 103–106.

Erlinge, S. (1981). Spacing-out systems and territorial behaviour in European otters. Abstract, Second International Otter Colloquium, Norwich, September 1981.

Erlinge, S. and Jensen, B. (1981). The diet of otters *Lutra lutra* L. in Denmark. *Nat. Jut.*, 19, 161–165.

Estes, J. A. (1980). *Enhydra lutris*. *Mammalian Species*, 133, 1–8.

Estes, J. A., Jameson, R. J. and Rhode, E. B. (1982). Activity and prey selection in the sea otter: influence of population status on community structure. *Am. Nat.*, 120, 242–258.

Estes, J. A. and Palmisano, J. F. (1974). Sea otters: their role in structuring nearshore communities. *Science*, 185, 1058–1060.

Estes, J. A. and Smith, N. S. (1973). Research on the sea otter, Amchitka Island, Alaska. *Natl. Tech. Inf. Serv., U.S. Dept. Commerce, Springfield, Virginia, NVO*, No. 520–1, 1–68.

Estes, J. A., Smith, N. S. and Palmisano, J. F. (1978). Sea otter predation and community organization in the western Aleutian Islands, Alaska. *Ecology*, 59, 822–833.

Ewer, R. F. (1973). *The carnivores*. Weidenfeld and Nicholson, London.

Fairley, J. S. (1972). Food of otters (*Lutra lutra*) from Co. Galway, Ireland and notes on aspects of their biology. *J. Zool., Lond.*, 166, 469–474.

Fairley, J. S. (1984). Otters feeding on breeding frogs. *Ir. Nat. J.*, 21, 372.

Fairley, J. S. and Wilson, S. C. (1972). Autumn food of otters (*Lutra lutra*) on the Agivey River, County Londonderry, Northern Ireland. *J. Zool., Lond.*, 166, 468–469.

Farney, J. P. and Jones, J. K. (1978). Recent records of the river otter from Nebraska. *Trans. Kansas Acad. Sci.*, 81, 275–276.

Ferrand de Almeida, F. (1980). Über das vorkommen des Fischotters in Portugal. In Reuther, C. and Festetics, A. (1980), *loc. cit.*, pp. 141–143.

Ferrand de Almedia, F., Ferrand de Almeida, P., Ferrand de Almeida, N. M., Moura, A. R., Da Graca Silva, M., Paiva, J. A. R., Nogueira, I. M., Soares, A. F., Pena dos Reis, R. P. B., Godinho, M. M. and Pereira de Oliveira, J. M. (1983). Aspectos faunísticos, florísticos, geólogicos e geográficos do Paúl de Arzila. *Ciênc. Biol. Ecol. Syst. (Portugal)*, 5, 43–78.

Field, R. J. (1970). Winter habits of the river otter (*Lutra canadensis*) in Michigan. *Michigan Academ.*, 3, 49–58.

Field, R. J. (1978). Observations on the presence of furbearers in the Sequatchie River, Tennessee. *J. Tennessee Acad. Sci.*, 53, 37–40.

Folio, M. R., Hennigan, C. H. and Errera, J. (1982). A comparison of five toxic metals among rural and urban children. *Environ. Pollut. Ser. A*, **29**, 261–269.

Foottit, R. G. and Butler, R. W. (1977). Predation on nesting glaucous-winged gulls by river otter. *Can. Field Nat.*, **91**, 189–190.

Förstner, U. and Wittmann, G. T. W. (1981). *Metal pollution in the aquatic environment.* Springer-Verlag, Berlin.

Foster-Turley, P., Davis, J. and Wright, C. (1983). The captive otter management project. *Otters, J. Otter Trust*, 1983, pp. 31–34.

Fowler, D., Cape, J. N., Leith, I. D., Paterson, I. S., Kinnaird, J. W. and Nicholson, I. A. (1982). Rainfall acidity in northern Britain. *Nature*, **297**, 383–386.

Frankel, D. H. and Soulé, M. E. (1981). *Conservation and evolution.* Cambridge University Press, Cambridge.

Friberg, L., Piscator, M., Nordberg, G. F. and Kjellström, T. (1974). *Cadmium in the environment.* 2nd edn. CRC Press, Ohio.

Furuya, Y. (1976). Otters in Padas Bay, Sabah, Malaysia. *J. Mammal Soc. Japan*, **7**, 34–43.

Gerell, R. (1967). Dispersal and acclimatization of the mink (*Mustela vison* Schreber) in Sweden. *Viltrevy*, **5**, 1–38.

Gilbert, F. F. and Nancekivell, E. G. (1982). Food habits of mink (*Mustela vison*) and otter (*Lutra canadensis*) in north-eastern Alberta. *Can. J. Zool.*, **60**, 1282–1288.

Gormally, M. J. and Fairley, J. S. (1982). Food of otters *Lutra lutra* in a freshwater lough and an adjacent brackish lough in the west of Ireland. *J. Zool., Lond.*, **186**, 463–474.

Gorman, M. L., Jenkins, D. and Harper, R. J. (1978). The anal scent sacs of the otter (*Lutra lutra*). *J. Zool., Lond.*, **186**, 463–474.

Gosling, L. M. (1982). A reassessment of the function of scent marking in territories. *Z. Tierpsychol.*, **60**, 89–118.

Grafen, A. (1984). Natural selection, kin selection and group selection. In Krebs, J. R. and Davies, N. B. (eds.) *Behavioural ecology: an evolutionary approach*, 2nd edn, pp. 62–84. Blackwell Scientific Publications, Oxford.

Green, J. (1977). Sensory perception in hunting otters, *Lutra lutra* L. *Otters, J. Otter Trust*, 1977, pp. 13–16.

Green, J. and Green, R. (1980). *Otter survey of Scotland 1977–79.* Vincent Wildlife Trust, London.

Green, J. and Green, R. (1981). The otter (*Lutra lutra* L.) in western France. *Mammal Rev.*, **11**, 181–187.

Green, J. and Green, R. (1985). Territoriality and home range in Scotland. In Duplaix, N. and Kempf, C. (1985), *loc. cit.*, in press.

Green, J., Green, R. and Jefferies, D. J. (1984). A radio-tracking survey of otters *Lutra lutra* on a Perthshire river system. *Lutra*, **27**, 85–145.

Greer, K. R. (1955). Yearly food habits of the river otter in the Thompson Lakes Region, north-western Montana, as indicated by scat analyses. *Am. Midl. Nat.*, **54**, 299–313.

Grenfell, W. E. (1978). Food habits of the river otter in Suisin Marsh, central California. *California–Nevada Wildl.*, 1978, pp. 65–73.

Griva, E. E. (1978). El programa de cría y preservación de *L. platensis* en Argentina. In Duplaix, N. (1978), *loc. cit.*, pp. 86–103.

Groupe Loutres (1983). Repartition de la loutre en France. *Bull. Groupe Loutres*, **15**, 17–21.

Gudger, E. W. (1927). Fishing with the otter. *Am. Nat.*, **61**, 193–225.

Halbrook, R. S., Jenkins, J. H., Bush, P. B. and Seabolt, N. D. (1981). Selected environmental contaminants in river otters (*Lutra canadensis*) of Georgia and their relationship to the posible decline of otters in North America. In Chapman, J. A. and Pursley, D. (eds.), *World Furbearer Conference Proceedings*, vol. III, pp. 1752–1762. World Furbearer Conference Inc., Frostburg, Maryland.

Hall, K. R. L. and Schaller, G. B. (1964). Tool using behaviour of the California sea otter. *J. Mammal.*, **45**, 287–298.

Hamilton, W. J. (1961). Late fall, winter and early spring foods of 141 otters from New York. *New York Fish Game J.*, **8**, 106–109.

Hamilton, W. J. and Eadie, W. R. (1964). Reproduction in the otter, *Lutra canadensis*. *J. Mammal.*, **45**, 242–252.

Harper, R. J. (1981). Sites of three otter (*Lutra lutra*) breeding holts in freshwater habitats. *J. Zool., Lond.*, **195**, 554–556.

Harper, R. J. and Jenkins, D. (1981). Mating behaviour in the European otter (*Lutra lutra*). *J. Zool., Lond.*, **195**, 556–558.

Harris, C. J. (1968). *Otters: a study of recent Lutrinae*. Weidenfeld and Nicholson, London.

Harris, M. P. (1984). *The puffin*. T. and A. D. Poyser, Calcot.

Harrison, B. (1984). Otters and fishing. *Institute of Fisheries Management News*, **14** (April 1984), 4–6.

Harrison, D. L. (1968). *The mammals of Arabia*, vol. 11, pp. 249–257. Ernest Benn Ltd, London.

Hartmann, J. (1977). Fischereiliche Veranderungen in kulturbedingt eutrophierenden Seen. *Schweiz. Z. Hydrol.*, **39**, 243–254.

Heggberget, T. M. and Myrberget, S. (1980). Der norwegische Fischotterbestand 1970–1977. In Reuther, C. and Festetics, A. (1980), *loc. cit.*, pp. 93–101.

Heidemann, G. (1980). Zur Lage des Fischotterbestandes in Schleswig-Holstein (Bundesrepublik Deutschland). In Reuther, C. and Festetics, A. (1980), *loc. cit.*, pp. 145–151.

Heidemann, G. (1981). Fischotter (*Lutra lutra* L.) in Schleswig-Holstein. *Zool. Anz.*, **207**, 210–221.

Hell, P. (1980). Die Situation des Fischotters in der ČSSR. In Reuther, C. and Festetics, A. (1980), *loc. cit.*, pp. 195–198.

Henny, C. J., Blus, L. J., Gregory, S. V. and Stafford, C. J. (1981). PCBs and organochlorine pesticides in wild mink and river otters from Oregon. In Chapman, J. A. and Pursley, D. (eds.), *World Furbearer Conference Proceedings*, vol. III, pp. 1763–1780. World Furbearer Conference Inc., Frostburg, Maryland.

Henshilwood, D. A. (1981). Status and diet of the otter (*Lutra lutra*) at Bosherston Lakes, Dyfed. Unpublished report to the Nature Conservancy Council, Aberystwyth.

Heptner, V. G. and Naumov, N. P. (eds.) (1974). *Die Säugetiere der Sowjetunion*. Gustav Fischer, Jena.

Herfst, M. S. (1984). Habitat and food of the otter *Lutra lutra* in Shetland. *Lutra*, 27, 57–70.

Hernandez, J. E. (1960). Contribucion al conocimiento de camarois de rio. *Pesca y Caza. Ministeria de Agricultura, Lima*, 10, 84–106.

Hershkovitz, P. (1969). The evolution of mammals on Southern continents. VI. The recent mammals of the neotropical region; a zoogeographic and ecological review. *Q. Rev. Biol.*, 44, 1–70.

Hewson, R. (1969). Couch building by otters (*Lutra lutra*). *J. Zool., Lond.*, 159, 524–527.

Hewson, R. (1973). Food and feeding habits of otters *Lutra lutra* at Loch Park, north-east Scotland. *J. Zool., Lond.*, 170, 143–162.

Hider, R. C., Mason, C. F. and Bakaj, M. E. (1982). *Chlorinated hydrocarbon pesticides and polychlorinated biphenyls in freshwater fishes in the United Kingdom, 1980–81*. Vincent Wildlife Trust, London.

Hill, E. P. and Lovett, J. W. (1975). Pesticide residues in beaver and river otter from Alabama. *Proc. 29th Ann. Conf. South-eastern Ass. Game Fish Commiss.*, 1975, pp. 365–369.

Hillegaart, V., Ostman, J. and Sandegren, F. (1981a). Area utilization and marking behaviour among two captive otter (*Lutra lutra* L.) pairs. Abstract, Second International Otter Colloquium, Norwich, September 1981.

Hillegaart, V., Ostman, J. and Sandegren, F. (1981b). Behavioural changes in captive female otters (*Lutra lutra* L.) around parturition. Abstract, Second International Otter Colloquium, Norwich, September 1981.

Hodl-Rohn, I. (1974). Verhaltensstudien an drei zahmen Glattottern, *Lutra (Lutrogale) perspicillata* (I. Geoffroy, 1826). *Säugetiere Mitt.*, 22, 17–28.

Hodl-Rohn, I. (1980). Die Situation des Fischotters im Bayerischen Wald (Bundesrepublik Deutschland). In Reuther, C. and Festetics, A. (1980), *loc. cit.*, pp. 175–177.

Holden, A. V. (1973). International co-operative study of organochlorine and mercury residues in wildlife, 1969–71. *Pestic. Monit. J.*, 7, 37–52.

Hönigsfeld, M. and Adamič, M. (1983). The otter (*Lutra lutra* L.) in Slovenia (Socialist Federative Republic of Yugoslavia). Unpublished report, Ljubljana.

Hornocker, M. G. (1969). Winter territoriality in mountain lions. *J. Wildl. Mgmt*, 33, 457–464.

Hornocker, M. G., Messick, J. P. and Melquist, W. E. (1983). Spatial strategies in three species of Mustelidae. *Acta Zool. Fenn.*, 174, 185–188.

Houk, J. L. and Geibel, J. J. (1974). Observation of underwater tool use by the sea otter, *Enhydra lutris* Linnaeus. *Calif. Fish Game*, 60, 207–208.

Humphrey, S. R. and Zinn, T. L. (1982). Seasonal habitat use by river otters and Everglades mink in Florida. *J. Wildl. Mgmt*, 46, 375–381.

Iverson, J. A. (1972). Basal energy metabolism of mustelids. *J. Comp. Physiol.*, 81, 341–344.

Jackson, W. L. and van Haveren, B. P. (1984). Design for a stable channel in coarse alluvium for riparian zone restoration. *Water Resources Bull*, 20, 695–703.

Jameson, R. J., Kenyon, K. W., Johnson, A. M. and Wight, H. M. (1982). History and status of translocated sea otter populations in North America. *Wildl. Soc. Bull.*, 10, 100–107.

Jarman, R. (1979). *Otter survey of the Somerset Levels, 1977–78*. Somerset Trust for Nature Conservation, Bridgwater.

Jarman, R. (1981). *Otters on Exmoor. A survey of otters in the Exmoor National Park*. Somerset Trust for Nature Conservation, Bridgwater, Somerset.

Jefferies, D. J. (1984). The otters return. *Nat. World*, 11, 10–12.

Jefferies, D. J., French, M. C. and Stebbings, R. E. (1974). Pollution and mammals. In *Monk's Wood Experimental Station Report for 1972–73*, pp. 13–15. N.E.R.C., Huntingdon.

Jefferies, D. J., Green, J. and Green, R. (1984). *Commercial fish and crustacean traps: a serious cause of otter* Lutra lutra (*L.*) *mortality in Britain and Europe*. Vincent Wildlife Trust, London.

Jefferies, D. J., Jessop, R. and Mitchell-Jones, A. J. (1983). Re-introduction of captive-bred European otters *Lutra lutra* to suitable field situations in England. *Otters, J. Otter Trust*, 1983, pp. 37–40.

Jefferies, D. J. and Mitchell-Jones, A. J. (1981). Preliminary research for a release programme for the European otter. *Otters, J. Otter Trust*, 1981, pp. 13–16.

Jefferies, D. J. and Wayre, P. (1983). Re-introduction to the wild of otters bred in captivity. *Otters, J. Otter Trust*, 1983, pp. 20–22.

Jenkins, A. L. (1982). *Otter distribution on the Western Cleddau*. Nature Conservancy Council/Vincent Wildlife Trust, Aberystwyth.

Jenkins, D. (1980). Ecology of otters in northern Scotland. 1. Otter (*Lutra*

lutra) breeding and dispersion in mid-Deeside, Aberdeenshire in 1974–79. *J. Anim. Ecol.*, **49**, 713–735.

Jenkins, D. (1981). Ecology of otters in northern Scotland. IV. A model scheme for otter *Lutra lutra* L. conservation in a freshwater system in Aberdeenshire. *Biol. Conserv.*, **20**, 123–132.

Jenkins, D. and Burrows, G. O. (1980). Ecology of otters in northern Scotland. III. The use of faeces as indicators of otter (*Lutra lutra*) density and distribution. *J. Anim. Ecol.*, **49**, 755–774.

Jenkins, D. and Harper, R. J. (1980). Ecology of otters in northern Scotland. II. Analyses of otter (*Lutra lutra*) and mink (*Mustela vison*) faeces from Deeside, N.E. Scotland in 1977–78. *J. Anim. Ecol.*, **49**, 737–754.

Jenkins, D., Walker, J. G. K. and McCowan, D. (1979). Analyses of otter (*Lutra lutra*) faeces from Deeside, N.E. Scotland. *J. Zool., Lond.*, **187**, 235–244.

Jenkins, J. H. (1983). The status and management of the river otter (*Lutra canadensis*) in North America. *Acta Zool. Fenn.*, **174**, 233–235.

Jensen, A. (1964). Odderen i Danmark. *Danske Vildtundersøgelser*, **11**, 1–48.

Jensen, B. (1980). Die situation des Fischotters in Dänemark. In Reuther, D. and Festetics, A. (1980), *loc. cit.*, pp. 115–116.

Jensen, S. (1972). The PCB story. *Ambio*, **1**, 123–131.

Jensen, S., Kihlström, J. E., Olsson, M., Lundberg, C. and Örberg, J. (1977). Effects of PCB and DDT on mink (*Mustela vison*) during the reproductive season. *Ambio*, **6**, 239.

Jernelov, A. and Martin, A. (1980). *Mercury in freshwater systems*. Swedish Water and Air Pollution Research Institute, Stockholm.

Johnson, A. M. (1982). Status of Alaska sea otter populations and developing conflicts with fisheries. *Trans. North Am. Wildl. Nat. Res. Conf.*, **47**, 283–299.

Johnstone, P. (1978). Breeding and rearing of the Canadian otter at Mole Hall Wildlife Park 1966–77. *Int. Zoo Yb.*, **18**, 143–147.

Johannson, I. (1961). Studies on the genetics of ranch bred mink. I. The results of an inbreeding experiment. *Z. Tierzücht. Züchtungs-Biol.*, **75**, 293–297.

Joint Otter Group (1977). *Otters 1977*. Nature Conservancy Council/Society for the Promotion of Nature Conservation, London.

Joint Otter Group (1979). *Otters 1979*. Society for the Promotion of Nature Conservation, Lincoln.

Kania, H. J. and O'Hara, J. (1974). Behavior alteration in a simple predator–prey system due to sub-lethal exposure to mercury. *Trans. Am. Fish Soc.*, **103**, 134–136.

Karr, J. R. and Schlosser, I. J. (1978). Water resources and the land–water interface. *Science*, **201**, 229–234.

Kaushik, N. K. and Hynes, H. B. N. (1971). The fate of dead leaves that fall into streams. *Arch. Hydrobiol.*, **68**, 465–515.

Kempf, C. (1981). Preliminary data on the status of the otter in France. Abstract, Second International Otter Colloquium, Norwich, September, 1981.

Kempf, C. (1985). The status of the otter in France. In Duplaix, N. and Kempf, C. (1985), *loc. cit.*, in press.

Kennedy, G. J. A., Cragg-Hine, D., Strange, C. D. and Stewart, D. A. (1983). The effects of a land drainage scheme on the salmonid populations of the River Camowen, Co. Tyrone. *Fish. Mgmt*, **14**, 1–16.

Kenyon, K. W. (1969). The sea otter in the Eastern Pacific Ocean. *U.S. Fish and Wildlife Service, North American Fauna*, no. 69.

Kenyon, K. W. (1981). Sea otter *Enhydra lutris* (Linnaeus, 1758). In Ridgway, S. H. and Harrison, R. J. (eds.), *Handbook of marine mammals*, vol. 1, pp. 209–223. Academic Press, London.

Kenyon, K. W. (1982). Sea otter. In Chapman, J. A. and Feldhamer, G. A. (eds.), *Wild mammals of North America*, pp. 704–710. Johns Hopkins University Press, Baltimore.

Kerr, S. R. and Werner, E. E. (1980). Niche theory in fisheries ecology. *Trans. Am. Fish. Soc.*, **109**, 254–260.

Kesteloot, E. (1980). Der Fischotter in Belgien. In Reuther, C. and Festetics, A. (1980), *loc. cit.*, pp. 135–137.

Khan, M. K. M. (1985). Travaux menés sur les loutres dans le sud-est de l'Asie. In Duplaix, N. and Kempf, C. (1985), *loc. cit.*, in press.

Kilham, L. (1982*a*). Common crows pulling the tail and stealing food from a river otter. *Florida Field Nat.*, **10**, 39–40.

Kilham, L. (1982*b*). Behavior of river otters by a water hole in a drought year. *Florida Field Nat.*, **10**, 60–61.

King, A. and Potter, A. (1980). *A guide to otter conservation for water authorities.* Vincent Wildlife Trust, London.

King, C. M. (1984). The origin and adaptive advantages of delayed implantation in *Mustela erminea*. *Oikos*, **42**, 126–128.

Kiss, I. K. (1985). Occurrence and feeding habits of otter, *Lutra lutra*, in Hungary. In Duplaix, N. and Kempf, C. (1985), *loc. cit.*, in press.

Kjellström, T., Borg, K. and Lind, B. (1978). Cadmium in feces as an estimator of daily cadmium intake in Sweden. *Environ. Res.*, **15**, 242–251.

Kleiman, D. G. (1980). The sociobiology of captive propagation. In Soulé, M. E. and Wilcox, B. A. (eds.), *Conservation biology: an evolutionary – ecological perspective*, pp. 243–262. Sinauer Associates, Sunderland, Mass.

Knudsen, G. J. and Hale, J. B. (1968). Food habits of otters in the Great Lakes region. *J. Wildl. Mgmt*, **32**, 89–93.

Kooyman, G. L., Davies, R. W. and Castellini, M. A. (1977). Thermal conductance of immersed pinniped and sea otter pelts before and after oiling with Prudhoe Bay crude. In Wolfe, D. A. (ed.), *Symposium on fate and effects of petroleum hydrocarbons in marine ecosystems and organisms*, pp. 151–157. Pergamon Press, New York.

Kraus, E. (1980). Probleme des Fischotterschutzes in Niederösterreich (Österreich). In Reuther, C. and Festetics, A. (1980), *loc. cit.*, pp. 205–210.

Krause, A. (1977). On the effect of marginal tree rows with respect to the management of small lowland streams. *Aquat. Bot.*, 3, 185–192.

Kruuk, H. (1978). Spatial organization and territorial behaviour of the European badger *Meles meles*. *J. Zool., Lond.*, 184, 1–19.

Kruuk, H., Gorman, M. and Parish, T. (1980). The use of ⁶⁵Zn for estimating populations of carnivores. *Oikos*, 34, 206–208.

Kruuk, H. and Hewson, R. (1978). Spacing and foraging of otters (*Lutra lutra*) in a marine habitat. *J. Zool., Lond.*, 185, 205–212.

Kruuk, H. and Parish, T. (1981). Feeding specialization of the European badger *Meles meles* in Scotland. *J. Anim. Ecol.*, 50, 773–788.

Kucera, E. (1983). Mink and otter as indicators of mercury in Manitoba waters. *Can. J. Zool.*, 61, 2250–2256.

Kučera, L. (1980). Bestandsentwicklung des Fischotters in Böhmen und Mähren (CSSR). In Reuther, C. and Festetics, A. (1980), *loc. cit.*, pp. 199–204.

Kucherenko, S. P. (1976). The common otter (*L. lutra*) in the Amur-Ussuri district. *Zool. Zh.*, 55, 904–911.

Lagler, K. F. and Ostenson, B. T. (1942). The early spring food of the otter in Michigan. *J. Wildl. Mgmt*, 6, 244–254.

Laidler, E. (1982). A study of the giant otter *Pteronura brasiliensis*. Movement patterns, territoriality, development and ecology in Guyana. *Otters, J. Otter Trust*, 1982, pp. 13–21.

Laidler, K. and Laidler, E. (1983). *The river wolf*. George Allen and Unwin, London.

Lawrence, M. J. and Brown, R. W. (1967). *Mammals of Britain. Their tracks, trails and signs*. Blandford Press, London.

Leith, D. E. (1976). Comparative mammalian respiratory mechanics. *Physiologist*, 19, 485–510.

Lensink, C. J. (1960). Status and distribution of sea otters in Alaska. *J. Mammal.*, 41, 172–182.

Lenton, E. (1985). The otter in the Somerset Levels. In Duplaix, N. and Kempf, C. (1985), *loc. cit.*, in press.

Lenton, E. J., Chanin, P. R. F. and Jefferies, D. J. (1980). *Otter survey of England 1977–79*. Nature Conservancy Council, London.

Leslie, G. (1970). Observations on the Oriental short-clawed otter, *Aonyx cinerea*, at Aberdeen Zoo. *Int. Zoo Yb.*, 10, 79–81.

Lever, C. (1977). *The naturalized animals of the British Isles*. Hutchinson, London.

Lever, C. (1978). The not so innocuous mink? *New Sci.*, **78**, 812–813.

Lewis, G. and Williams, G. (1984). *Rivers and wildlife handbook: a guide to practices which further the conservation of wildlife on rivers*. Royal Society for the Protection of Birds, Sandy, and Royal Society for Nature Conservation, Lincoln.

Leyhausen, P. (1971). Dominance and territoriality as complemented in mammalian social structure. In Esser, A. H. (ed.), *Behavior and environment: the use of space by animals and men*, pp. 22–33. Plenum Press, New York.

Libois, R. M., Philippart, J. C., Rosoux, R. and Vranken, M. (1982). Quel avenir pour la loutre en Belgique? *Cah. Ethol. appl.*, **2**, 1–15.

Liers, E. E. (1960). Notes on breeding the Canadian otter. *Int. Zoo Yb.*, **2**, 84–85.

Lightfoot, A. (1981). Coastal otters in Norway. Abstract, Second International Otter Colloquium, Norwich, September 1981.

Liles, G. and Jenkins, L. (1984). A field survey for otters (*Lutra lutra*) in Yugoslavia. *J. Zool., Lond.*, **203**, 282–284.

Limbaugh, C. (1961). Observations on the California sea otter. *J. Mammal.*, **42**, 271–273.

Lloyd, J. L. (1962). Where are the otters? *Gamekeeper and Countryside*, **65**, 299–300.

Loranger, A. J. (1981). Late fall and early winter foods of the river otter (*Lutra canadensis*) in Massachusetts, 1976–1978. In Chapman, J. A. and Pursley, D. (eds.), *World Furbearer Conference Proceedings*, vol. 1, pp. 599–605. World Furbearer Conference Inc., Frostburg, Maryland.

Loughlin, T. R. (1980a). Radio-telemetric determination of the 24-hour feeding activities of sea otters, *Enhydra lutris*. In Amlaner, C. J. and Macdonald, D. W. (eds.), *A handbook of biotelemetry and radio-tracking*, pp. 717–723. Pergamon, Oxford.

Loughlin, T. R. (1980b). Home range and territoriality of sea otters near Monterey, California. *J. Wildl. Mgmt*, **44**, 576–582.

Loughlin, T. R., Ames, J. A. and Vandevere, J. E. (1981). Annual reproduction, dependency period and apparent gestation period in two California sea otters, *Enhydra lutris*. *Fish. Bull.*, **79**, 347–349.

Love, J. A. (1983). *The return of the sea eagle*. Cambridge University Press, Cambridge.

Luahachinda, V. and Hill, E. P. (1978). Winter food habits of river otters from Alabama and Georgia. *Proc. Ann. Conf. S. E. Assoc. Fish and Wildlife Agencies*, **31**, 246–253.

Lydekker, R. (1896). *A handbook to the British Mammalia*. Edward Lloyd, London.

McCleneghan, K. and Ames, J. A. (1976). A unique method of prey capture by a sea otter *Enhydra lutris*. *J. Mammal.*, **57**, 401–412.

Macdonald, D. W. (1980). Patterns of scent marking with urine and faeces amongst carnivore communities. *Symp. Zool. Soc. Lond.*, **45**, 107–139.

Macdonald, S. M. (1983). The status of the otter (*Lutra lutra*) in the British Isles. *Mammal Rev.*, **13**, 11–23.

Macdonald, S. and Duplaix, N. (1983). The otter, symbol of our threatened fauna. *Naturopa*, **45**, 14–19.

Macdonald, S. and Mason, C. F. (1976). The status of the otter (*Lutra lutra* L.) in Norfolk. *Biol. Conserv.*, **9**, 119–124.

Macdonald, S. M. and Mason, C. F. (1980). Observations on the marking behaviour of a coastal population of otters. *Acta Theriol.*, **25**, 245–253.

Macdonald, S. M. and Mason, C. F. (1982*a*). Otters in Greece. *Oryx*, **16**, 240–244.

Macdonald, S. M. and Mason, C. F. (1982*b*). The otter *Lutra lutra* in central Portugal. *Biol. Conserv.*, **22**, 207–215.

Macdonald, S. M. and Mason, C. F. (1983*a*). The otter *Lutra lutra* in southern Italy. *Biol. Conserv.*, **25**, 95–101.

Macdonald, S. M. and Mason, C. F. (1983*b*). The otter (*Lutra lutra*) in Tunisia. *Mammal. Rev.*, **13**, 35–37.

Macdonald, S. M. and Mason, C. F. (1983*c*). Some factors influencing the distribution of otters (*Lutra lutra*). *Mammal Rev.*, **13**, 1–10.

Macdonald, S. M. and Mason, C. F. (1984). Otters in Morocco. *Oryx*, **18**, 157–159.

Macdonald, S. M. and Mason, C. F. (1985). Otters, their habitat and conservation in north-east Greece. *Biol. Conserv.*, **31**, 191–210.

Macdonald, S. M., Mason, C. F. and Coghill, I. S. (1978). The otter and its conservation in the River Teme catchment. *J. Appl. Ecol.*, **15**, 373–384.

Macdonald, S. M., Mason, C. F. and De Smet, K. (1985). The otter (*Lutra lutra*) in north-central Algeria. *Mammalia*, in press.

McEwen, F. L. and Stephenson, G. R. (1979). *The use and significance of pesticides in the environment*. Wiley, New York.

McFadden, Y. M. T. and Fairley, J. S. (1984). Food of otters *Lutra lutra* (L.) in an Irish limestone river system with special reference to the crayfish *Austrapotamobius pallipes* (Lereboullet). *J. Life Sci. R. Dubl. Soc.*, **5**, 65–76.

Magoun, A. J. and Valkenburg, P. (1977). The river otter (*Lutra canadensis*) on the north slope of the Brook's Range, Alaska. *Can. Field-Nat.*, **91**, 303–305.

Mann, K. H. (1969). The dynamics of aquatic ecosystems. *Adv. Ecol. Res.*, **6**, 1–71.

Mason, C. F. (1981). *Biology of freshwater pollution*. Longman, London.

Mason, C. F. (1985). The effects of pollutants on otters. In Kempf, C. and
Duplaix, N. (1985), *loc. cit.*, in press.

Mason, C. F. and Macdonald, S. M. (1980). The winter diet of otters (*Lutra
lutra*) on a Scottish sea loch. *J. Zool., Lond.*, **192**, 558–561.

Mason, C. F. and Macdonald, S. M. (1982). The input of terrestrial
invertebrates from tree canopies to a stream. *Freshwater Biol.*, **12**, 305–311.

Mason, C. F. and Macdonald, S. M. (1983). Some factors influencing the
distribution of mink (*Mustela vison*). *J. Zool., Lond.*, **200**, 281–283.

Mason, C. F., Macdonald, S. M. and Aspden, V. J. (1982). *Metals in
freshwater fishes in the United Kingdom, 1980–1981.* Vincent Wildlife
Trust, London.

Mason, C. F., Macdonald, S. M. and Hussey, A. (1984). Structure,
management and conservation value of the riparian woody plant
community. *Biol. Conserv.*, **29**, 201–216.

Medway, Lord (1978). *The wild mammals of Malaya (Peninsular Malaysia)
and Singapore*, 2nd edn. Oxford University Press, Oxford.

Meehan, W. R., Swanson, F. J. and Sedell, J. R. (1977). Influences of
riparian vegetation on aquatic ecosystems with particular reference to
salmonid fishes and their food supply. In Johnson, R. R. and Jones, D. A.
(eds.), *Importance, preservation and management of riparian habitat: a
symposium*, pp. 137–145. U.S.D.A. Forest Service, Fort Collins, Colorado.

Melquist, W. E. (1984). *Status survey of otters (Lutrinae) and spotted cats
(Felidae) in Latin America.* Report to I.U.C.N. College of Forestry,
Wildlife and Range Sciences, University of Idaho.

Melquist, W. E. and Hornocker, M. G. (1979). Methods and techniques for
studying and censusing river otter populations. *Univ. Idaho For. Wildl.
and Range Exp. Stn Tech. Rep.* 8.

Melquist, W. E. and Hornocker, M. G. (1983). Ecology of river otters in
west central Idaho. *Wildl. Monogr.*, **83**, 1–60.

Melquist, W. E., Whitman, J. S. and Hornocker, M. G. (1981). Resource
partitioning and coexistence of sympatric mink and river otter
populations. In Chapman, J. A. and Pursley, D. (eds.), *World Furbearer
Conference Proceedings*, vol. 1, pp. 187–220. World Furbearer Conference
Inc., Frostburg, Maryland.

Melville, D. S. and Horton, B. (1983). *Mai Po Marshes.* World Wildlife
Fund, Hong Kong.

Mikuriya, M. (1976). Notes on the Japanese otter, *Lutra lutra whiteleyi*
(Gray). *J. Mammal. Soc. Japan*, **6**, 214–217.

Miles, H. (1984). *The track of the wild otter.* Elm Tree Books, London.

Miller, S. D., Rottmann, J., Raedeke, K. J. and Taber, R. D. (1983).
Endangered mammals of Chile: status and conservation. *Biol. Conserv.*,
25, 335–352.

Milner, N. J., Gee, A. S. and Hemsworth, R. J. (1978). The production of brown trout, *Salmo trutta*, on tributaries of the Upper Wye, Wales, *J. Fish Biol.*, 13, 599–612.

Mitchell-Jones, A. J., Jefferies, D. J., Twelves, J., Green, J. and Green, R. (1984). A practical system of tracking otters *Lutra lutra* using radio-telemetry and 65-Zn. *Lutra*, 27, 71–74.

Modafferi, R. and Yocom, C. F. (1980). Summer food of river otter in north coastal California lakes. *Murrelet*, 61, 38–41.

Moors, P. J. (1980). Sexual dimorphism in the body size of mustelids (Carnivora): the roles of food habits and breeding systems. *Oikos*, 34, 147–158.

Morejohn, G. V. (1969). Evidence of river otter feeding on freshwater mussels and range extension. *Calif. Fish Game*, 55, 83–85.

Morris, R., Ellis, D. V. and Emerson, B. P. (1981). The British Columbia transplant of sea otters *Enhydra lutris*. *Biol. Conserv.*, 20, 291–295.

Morrison, P., Rosenmann, M. and Estes, J. A. (1974). Metabolism and thermoregulation in the sea otter. *Physiol. Zool.*, 47, 218–229.

Mortimer, M. A. E. (1963). Notes on the biology and behaviour of the spotted-necked otter (*Lutra maculicollis*). *Puku*, 1, 192–206.

Mowbray, E. E., Pursley, D. and Chapman, J. A. (1979). The status, population characteristics and harvest of the river otter in Maryland. *Publ. Wildl. Ecol.* no. 2, Md Wildl. Admin.

Müffling, S. von (1977). *Fischotter in Europa*. Loizenkirchen.

Müller, H.-U., Martin, C. and Diethelm, P. (1976). *La loutre. Sa présence, ses conditions d'existence, sa conservation en Suisse*. Studies of the Zoological Institute, University of Zurich, Zurich.

Myers, N. (1979). *The sinking ark*. Pergamon Press, Oxford.

Myers, N. (1983). *A wealth of wild species*. Westview Press, Boulder, Colorado.

Myrberget, S. and Fröiland, O. (1972). Oteren in Norge omkring 1970. *Fauna*, 25, 149–159.

National Academy of Sciences, National Academy of Engineering (1973). Section III – Freshwater aquatic life and wildlife, Water Quality Criteria. *Ecological Research Series*, EPA-R3-73-033, March 1973, pp. 106–113.

National Water Council (1981). *River quality; the 1980 survey and future outlook*. N.W.C., London.

Nature Conservancy Council (1984). *Nature conservation in Britain*. N.C.C., Shrewsbury.

Nechay, G. (1980). Die situation des Fischotters in Ungarn. In Reuther, C. and Festetics, A. (1980), *loc. cit.*, pp. 215–221.

Newby, T. C. (1975). A sea otter (*Enhydra lutris*) food dive record. *Murrelet*, 56, 7.

Newson, M. (1984). River processes and form. In Lewis and Williams (1984), *loc. cit.*, pp. 3–9.

222   REFERENCES

Newton, I. and Haas, M. B. (1984). The return of the sparrowhawk. *Brit. Birds*, 77, 47–70.

Nicholson, J. K., Kendall, M. D. and Osborn, D. (1983). Cadmium and mercury nephrotoxicity. *Nature*, 304, 633–635.

Nilsson, G. (1980). River otter pelt exports. *Brightwater Journal*, 1, 5.

Novakova, E. (1985). The otter in Czechoslovakia. In Duplaix, N. and Kempf, C. (1985), *loc. cit.*, in press.

Novikov, G. A. (1956). Carnivorous mammals of the fauna of the USSR. Keys to the fauna of the USSR. *Zool. Inst. Acad. Sci. USSR*, 62, 213–218.

O'Connor, D. J. and Nielsen, S. W. (1981). Environmental survey of methyl-mercury levels in the wild mink (*Mustela vison*) and otter (*Lutra canadensis*) from the north-eastern United States and experimental pathology of methylmercurialism in the otter. In Chapman, J. A. and Pursley, D. (eds.), *World Furbearer Conference Proceedings*, vol. III, pp. 1728–1745. World Furbearer Conference Inc., Frostburg, Maryland.

Odemar, N. W. and Wilson, K. C. (1969). Results of sea otter capture, tagging and transporting operations by the California Department of Fish and Game. *Proc. Ann. Conf. Biol. Sonar and Diving Mammals*, 6, 73–79.

Olsson, M., Reutergårdh, L. and Sandegren, F. (1981). Var är uttern? *Sveriges Natur*, 6/8, 234–240.

Olsson, M. and Sandegren, F. (1985). Otters in Sweden. In Kempf, C. and Duplaix, N. (1985), *loc. cit.*, in press.

Ormond, R. (1981). *Sir Edwin Landseer*. Philadelphia Museum of Art, Philadelphia.

Ostfield, R. S. (1982). Foraging strategies and prey switching in the California sea otter. *Oecologia*, 53, 170–178.

Palmisano, J. F. (1983). Sea otter predation: its role in structuring rocky intertidal communities in the Aleutian Islands, Alaska, U.S.A. *Acta Zool. Fenn.*, 174, 209–211.

Pechlaner, H. and Thaler, E. (1983). Beitrag zur Fortpflanzungsbiologie des europäischen Fischotters (*Lutra lutra* L.). *Zool. Garten N.F., Jena*, 53, 49–58.

Pielowski, Z. (1980). Die Situation des Fischotters in Polen. In Reuther, C. and Festetics, A. (1980), *loc. cit.*, pp. 183–185.

Piotrowski, J. K. and Inskip, M. J. (1981). *Health effects of mercury*. M.A.R.C. Report no. 24, Chelsea College, London.

Poche, R. (1973). Niger's threatened Park W. *Oryx*, 12, 216–222.

Powell, R. A. (1979). Mustelid spacing patterns: variations on a theme by *Mustela*. *Z. Tierpsychol.*, 50, 153–165.

Prestt, I. (1984). Comment. *Birds*, 10(2), 4.

Prestt, I., Jefferies, D. J. and Moore, N. W. (1970). Polychlorinated biphenyls in wild birds in Britain and their avian toxicity. *Environ. Pollut.*, 1, 3–26.

Prigioni, C. (1983). Confermata la presenza della lontra (*Lutra lutra*) nel Lago di Mezzola (Lombardia). *Natura-Soc. ital. Sci. nat.*, *Museo civ. stor. nat. e Acquario civ.*, *Milano*, **74**, 125–126.

Procter, J. (1963). A contribution to the natural history of the spotted-necked otter (*Lutra maculicollis* Lichtenstein) in Tanganyika. *E. Afr. Wildl. J.*, **1**, 93–102.

Pulliainen, E. (1985). Recent decline in otters in Finland. In Duplaix, N. and Kempf, C. (1985), *loc. cit.*, in press.

Ralls, K., Ballou, J. and Brownell, R. L. (1983). Genetic diversity in California sea otters: theoretical considerations and management implications. *Biol. Conserv.*, **25**, 209–232.

Ratcliffe, D. A. (1980). *The peregrine falcon*. Poyser, Calton.

Reuther, C. (1980*a*). Der Fischotter, *Lutra lutra* L. in Niedersachsen. *Natursch. Landschaftspf. Niedersachsen*, **11**, 1–182.

Reuther, C. (1980*b*). Zur Situation des Fischotters in Europa. In Reuther, C. and Festetics, A. (1980), *loc. cit.*, pp. 71–92.

Reuther, C. (1981). Preliminary findings of a survey of otters in captivity. Abstract, Second International Otter Colloquium, Norwich, 1981.

Reuther, C. (1984). Über die Arbeit der 'Aktion Fischotterschutz e.V'. *Lutra*, **27**, 45–52.

Reuther, C. and Festetics, A. (eds.) (1980). *Der Fischotter in Europa-Verbreitung, Bedrohung, Erhaltung*. Selbstverlag, Oderhaus and Göttingen.

Ribic, C. A. (1982*a*). Autumn activity of sea otters in California. *J. Mammal.*, **63**, 702–706.

Ribic, C. A. (1982*b*). Autumn movement and home range of sea otters in California. *J. Wildl. Mgmt*, **46**, 795–801.

Rickard, D. G. and Dulley, M. E. R. (1983). The levels of some heavy metals and chlorinated hydrocarbons in fish from the tidal Thames. *Environ. Pollut. Ser. B*, **5**, 101–119.

Rowbottom, M. J. and Rowbottom, C. H. (1980). Watching otters in the West Highlands of Scotland. *Jotters*, **6**, 2.

Rowe-Rowe, D. T. (1977*a*). Prey capture and feeding behaviour of South African otters. *Lammergeyer*, **23**, 13–21.

Rowe-Rowe, D. T. (1977*b*). Variations in the predatory behaviour of clawless otter. *Lammergeyer*, **23**, 22–27.

Rowe-Rowe, D. T. (1977*c*). Food ecology of otters in Natal, South Africa. *Oikos*, **28**, 210–219.

Rowe-Rowe, D. T. (1978*a*). The small carnivores of Natal. *Lammergeyer*, **25**, 1–48.

Rowe-Rowe, D. T. (1978*b*). Biology of the two otters in South Africa. In Duplaix, N. (1978), *loc. cit.*, pp. 130–139.

Rowe-Rowe, D. (1985). Distribution and status of two otter species in some South African countries. In Duplaix, N. and Kempf, C. (1985), *loc. cit.*, in press.

Royal Society for the Protection of Birds (1978*a*). *A survey of the birds of the River Wye.* R.S.P.B., Sandy, Beds.

Royal Society for the Protection of Birds (1978*b*). *A survey of the birds of the River Severn.* R.S.P.B., Sandy, Beds.

Royal Society for the Protection of Birds (1979). *A survey of the birds of the Rivers Avon (Hampshire/Wiltshire) and Stour (Dorset).* R.S.P.B., Sandy, Beds.

Royal Society for the Protection of birds (1983). *Land drainage in England and Wales: an interim report.* R.S.P.B., Sandy Beds.

Rüppell, G. and Rüppell, G.-M. (1980). Überfall eines Fingerotters, *Aonyx capensis,* auf einen Zwergflamingo, *Phoeniconaias minor. Zool. Gart.,* **50**, 274–275.

Ryder, R. A. (1955). Fish predation by the otter in Michigan. *J. Wildl. Mgmt,* **19**, 497–498.

Sandegren, F. (1981). Project otter – Sweden. Abstract, Second International Otter Colloquium, Norwich.

Sandegren, F. E., Chu, E. W. and Vandevere, J. E. (1973). Maternal behavior in the California sea otter. *J. Mammal.,* **54**, 668–679.

Sandegren, F., Olsson, M. and Reutergårdh, L. (1980). Der Rückgang der Fischotterpopulation in Schweden. In Reuther, C. and Festetics, A. (1980), *loc. cit.,* pp. 107–113.

Sandell, M. (1984). To have or not to have delayed implantation: the example of the weasel and the stoat. *Oikos,* **42**, 123–126.

Santos Reis, M. (1983). Status and distribution of the Portuguese mustelids. *Acta. Zool. Fenn.,* **174**, 213–216.

Schmitt, C. J., Ludke, J. L. and Walsh, D. F. (1981). Organochlorine residues in fish: national pesticide monitoring program, 1970–74. *Pestic. Monit. J.,* **14**, 136–155.

Schneider, K. B. (1972). Reproduction in the female sea otter. *Alaska Dept. Fish Game, Proj. Prog. Rep. Fed. Aid in Wildl. Restoration, Proj. W-17-4.*

Schneider, K. B. (1978). Sex and segregation in sea otters. *Alaska Dept. Fish Game, Final Rep., Fed. Aid. in Wildl. Restoration, Proj. W-17-4 to W-17-8.*

Schneider, K. B. and Faro, J. B. (1975). Effects of sea ice on sea otters (*Enhydra lutris*). *J. Mammal.,* **56**, 91–101.

Scott, T. G., Willis, Y. L. and Ellis, J. A. (1959). Some effects of a field application of dieldrin on wildlife. *J. Wild. Mgmt,* **23**, 409–427.

Shaw, S. B. (1971). Chlorinated hydrocarbon pesticides in California sea otters and harbor seals. *Calif. Fish Game,* **57**, 290–294.

Sheffy, T. B. and St Amant, J. R. (1982). Mercury burdens in furbearers in Wisconsin. *J. Wildl. Mgmt,* **46**, 1117–1120.

Sheldon, W. G. and Toll, W. G. (1964). Feeding habits of the river otter in a reservoir in central Massachusetts. *J. Mammal.*, 45, 449–455.

Sherrod, S. K., Estes, J. A. and White, C. M. (1975). Depredation of sea otter pups by bald eagles at Amchitka Island, Alaska. *J. Mammal.*, 56, 701–703.

Shimek, S. J. (1977). The underwater foraging habits of the sea otter, *Enhydra lutris*. *Calif. Fish Game*, 63, 120–122.

Shimek, S. J. and Monk, A. (1977). The daily activity of the sea otter off the Monterey Peninsula, California. *J. Wildl. Mgmt*, 41, 277–283.

Sielfeld, W., Venegas, C. and Atalah, A. (1977). Consideraciones acerca del estado de los mamíferos marinos en Chile. *An. Inst. Patagonia*, 8, 297–315.

Sikora, S. (1984). Wystepowanie wydry *Lutra lutra* (L.) w Polsce. *Pr. Kom. Nauk roln. leśn, Poznan*, 57, 253–268.

Sikora, S. (1985). The otter in Poland. In Duplaix, N. and Kempf, C. (1985), *loc. cit.*, in press.

Simenstad, C. A., Estes, J. A. and Kenyon, K. W. (1978). Aleuts, sea otters, and alternate stable-state communities. *Science*, 200, 403–411.

Simões, P. (1977–1982). Uma população de lontras do litoral Português. *Bolm. Liga Prot. Nat.*, 16, 17–19.

Simões Graça, M. A. and Ferrand de Almeida, F. X. (1983). Contribuição para o conhecimento da lontra (*Lutra lutra* L.) num secto da bacia do Rio Mondego. *Ciênc. Biol. Ecol. Syst.*, 5, 33–42.

Siniff, D. B., Williams, T. D., Johnson, A. M. and Garshelis, D. L. (1982). Experiments on the response of sea otters *Enhydra lutris* to oil contamination. *Biol. Conserv.*, 23, 261–272.

Sly, J. M. A. (1977). *Review of usage of pesticides in agriculture and horticulture in England and Wales, 1965–1974.* Survey Report 8. Ministry of Agriculture, Fisheries and Food, Pinner.

Sly, J. M. A. (1981). *Review of usage of pesticides in agriculture, horticulture and forestry in England and Wales, 1975–79. Survey Report 23.* Ministry of Agriculture, Fisheries and Food, Pinner.

Smith, B. D. (1980). The effects of afforestation on the trout of a small stream in southern Scotland. *Fish. Mgmt*, 11, 39–58.

Smith, G. J. and Rongstad, O. J. (1981). Heavy metals in mammals from two unmined copper–zinc deposits in Wisconsin. *Bull. Environ. Contam. Toxicol.*, 27, 28–33.

Smith, N. J. H. (1981). Caimans, capybaras, otters, manatees and man in Amazonia. *Biol. Conserv.*, 19, 177–187.

Society for the Promotion of Nature Conservation (undated). *Artificial otter holts.* S.P.N.C., Lincoln.

Spiridonov, G. (1985). The otter in Bulgaria. In Duplaix, N. and Kempf, C. (1985), *loc. cit.*, in press.

Stenson, G. B., Badgero, G. A. and Fisher, H. D. (1983). Food habits of the river otter *Lutra canadensis* in the marine environment of British Columbia. *Can. J. Zool.*, **62**, 88–91.

Stephens, M. N. (1957). *The otter report.* University Federation for Animal Welfare, Potters Bar.

Stoddart, D. M. (1980). *The ecology of vertebrate olfaction.* Chapman and Hall, London.

Strickland, A. H. (1966). Some estimates of insecticide and fungicide usage in agriculture and horticulture in England and Wales, 1960–64. *J. Appl. Ecol.*, *3* (suppl.), 3–13.

Stubbe, M. (1969). Zür Biologie und zum Schutz des Fischotters *Lutra lutra* (L.). *Arch. Naturschutz Landschaftsforsch.*, 9, 315–324.

Stubbe, M. (1970). Zur Evolution der analen Markierungsorgane bei Musteliden. *Biol. Zbl.*, 89, 213–223.

Stubbe, M. (1977). Der Fischotter *Lutra lutra* (L. 1758) in der DDR. *Zool. Anz.*, 199, 265–285.

Stubbe, M. (1980). Die Situation des Fischotters in der D.D.R. In Reuther, C. and Festetics, A. (1980), *loc. cit.*, pp. 179–182.

Stubbe, M. (1985*a*). Status of the otter in the D.D.R. In Duplaix, N. and Kempf, C. (1985), *loc. cit.*, in press.

Stubbe, M. (1985*b*). The status of the otter in Mongolia. In Duplaix, N. and Kempf, C. (1985), *loc. cit.*, in press.

Sullivan, J. F., Atchison, G. J., Kolar, D. J. and McIntosh, A. W. (1978). Changes in the predator–prey behaviour of fathead minnows (*Pimephales promelas*) and largemouth bass (*Micropterus salmoides*) caused by cadmium. *J. Fish. Res. Bd Canada*, 35, 446–451.

Swales, S. (1982). A 'before and after' study of the effects of land drainage works on fish stocks in the upper reaches of a lowland river. *Fish. Mgmt*, 13, 105–114.

Swedish Ministry of Agriculture (1983). *Acidification – a boundless threat to our environment.* Solna, Sweden.

Tabor, J. E. and Wight, H. M. (1977). Population status of river otter in western Oregon. *J. Wildl. Mgmt*, 41, 692–699.

Tankó, I. and Tassi, I. (1978). A vidra életmódjaról és halászati kártételéröl. *Halaszat*, 24, 72–75.

Tayler, C. (1970). The elusive otter. *Eastern Cape Naturalist*, 39, 20–23.

Thesiger, W. (1964). *The marsh arabs.* Longman, London.

Thornback, J. and Jenkins, M. (1982). *The IUCN mammal red data book, part 1.* I.U.C.N., Gland, Switzerland.

Tinelli, P. and Tinelli, A. (1980). The otter in Europe. Italy. *Jotters*, no. 6. Society for the Promotion of Nature Conservation, Lincoln.

Tomlinson, D. N. S. (1974). Studies of the purple heron, part 1: heronry structure, nesting habits and reproductive success. *Ostrich*, 45, 175–181.

Toweill, D. E. (1974). Winter food habits of river otters in western Oregon. *J. Wildl. Mgmt*, **38**, 107–111.

Toweill, D. E. and Tabor, J. E. (1982). River otter. In Chapman, J. A. and Feldhamer, G. A. (eds.), *Wild mammals of North America*, pp. 688–703. Johns Hopkins University Press, Baltimore.

Trowbridge, B. J. (1983). Olfactory communication in the European otter (*Lutra l. lutra*). Ph.D. thesis, University of Aberdeen.

Tumlison, C. R., King, A. W. and Johnston, L. (1981). The river otter in Arkansas: 1. Distribution and harvest trends. *Arkansas Acad. Sci. Proc.*, **35**, 74–77.

Turtle, E. E., Taylor, A., Wright, E. N., Thearle, R. J. P., Egan, H., Evans, W. H. and Soutar, N. M. (1963). The effects on birds of certain chlorinated insecticides used on seed dressings. *J. Sci. Food Agric.*, **14**, 456–477.

Twelves, J. (1982). A record of the body measurements and weights of otters in Uist. *Hebridean Naturalist, 1982*, 3 pp.

Twelves, J. (1983). Otter (*Lutra lutra*) mortalities in lobster creels. *J. Zool., Lond.*, **201**, 585–588.

Van Blaricom, G. R. and Jameson, R. J. (1982). Lumber spill in central California waters: implications for oil spills and sea otters. *Science*, **215**, 1503–1505.

Van der Zee, D. (1981). Prey of the Cape clawless otter (*Aonyx capensis*) in the Tsitsikama Coastal National Park, South Africa. *J. Zool., Lond.*, **194**, 467–483.

Van der Zee, D. (1982). Density of Cape clawless otters *Aonyx capensis* (Schinz, 1821) in the Tsitsikama Coastal National Park. *S. Afr. J. Wildl. Res.*, **12**, 8–13.

Vandevere, J. E. (1970). Reproduction in the southern sea otter. *Proc. Ann. Conf. Biol. Sonar and Diving Mammals*, **7**, 221–227.

Vandevere, J. E. (1983). Annual reproduction and gestation periods of one wild southern sea otter (*Enhydra lutris nereis*). *Otters, J. Otter Trust*, 1983, pp. 28–30.

Van Wagenen, R. F., Foster, M. S. and Burns, F. (1981). Sea otter predation on birds near Monterey, California. *J. Mammal.*, **62**, 433–434.

van Wijngaarden, A. (1980). Der Status des Fischotters in den Niederlanden. In Reuther, C. and Festetics, A. (1980), *loc. cit.*, pp. 123–128.

van Wijngaarden, A. and van de Peppel, J. (1970). De otter, *Lutra lutra* (L.) in Nederland. *Lutra*, **12**, 3–70.

van Zyll de Jong, C. G. (1972). A systematic review of the Nearctic and Neotropical river otters (Genus *Lutra*, Mustelidae, Carnivora). *Life Sci. Contrib., R. Ontario Mus.*, **80**, 1–104.

Veen, J. (1984). De verspreiding en enkele oecologische aspecten van de otter *Lutra lutra* in Nederland. *Lutra*, **27**, 25–35.

Waldron, H. A. (1980). Lead. In Waldron, H. A. (ed.), *Metals in the environment*, pp. 155–197. Academic Press, London.

Walton, I. (1653). *The compleat angler*. Richard Marriot, London.

Watson, H. (1978). *Coastal otters* (Lutra lutra L.) *in Shetland*. Vincent Wildlife Trust, London.

Wayre, P. (1974). Otters in western Malaysia. *Otter Trust Annual Report*, 1974, pp. 16–38.

Wayre, P. (1978). Status of otters in Malaysia, Sri Lanka and Italy. In Duplaix, N. (1978), *loc. cit.*, pp. 152–155.

Wayre, P. (1979a). *The private life of the otter*. Batsford, London.

Wayre, P. (1979b). Otter havens in Norfolk and Suffolk, England. *Biol. Conserv.*, 16, 73–81.

Webb, J. B. (1975). Food of the otter (*Lutra lutra*) on the Somerset Levels. *J. Zool., Lond.*, 177, 486–491.

Webb, J. (1976). *Otter spraint analysis*. The Mammal Society, Reading.

Weir, V. and Bannister, K. E. (1973). The food of the otter in the Blakeney area. *Trans. Norf. Norw. Nats. Soc.*, 22, 377–382.

Weir, V. and Bannister, K. E. (1977). Additional notes on the food of the otter in the Blakeney area. *Trans. Norf. Norw. Nats. Soc.*, 24, 85–88.

West, R. B. (1975). The Suffolk otter survey. *Suffolk Nat. Hist.*, 16, 378–388.

Westlake, G. E., Bunyan, P. J. and Stanley, P. I. (1978). Variation in the response of plasma enzyme activities in avian species dosed with carbophenothion. *Ecotoxicol. Environ. Safety*, 2, 151–159.

Wild, P. W. and Ames, J. A. (1974). A report on the sea otter, *Enhydra lutris*, L., in California. *California Dept. Fish Game, Mar. Res. Tech. Rep.* 20, 1–93.

Williamson, K. (1971). A bird census of a Dorset dairy farm. *Bird Study*, 18, 80–96.

Wilson, A. (1969). *Further review of certain persistent organochlorine pesticides used in Great Britain*. H.M.S.O., London.

Winneke, G., Brockhaus, A. and Baltissen, R. (1977). Neuro-behavioural and systemic effects of long term blood lead elevation in rats. *Arch. Toxicol.*, 37, 247–263.

Wise, M. H. (1980). The use of fish vertebrae in scats for estimating prey size of otters and mink. *J. Zool., Lond.*, 192, 25–31.

Wise, M. H., Linn, I. J. and Kennedy, C. R. (1981). A comparison of the feeding biology of mink *Mustela vison* and otter *Lutra lutra*. *J. Zool., Lond.*, 195, 181–213.

Wlodek, K. (1980). Der Fischotter in der Provinz Pomorze Zachodnie (West-Pommern) in Polen. In Reuther, C. and Festetics, A. (1980), *loc. cit.*, pp. 187–194.

Wlodek, K. (1985). Causes of the decline of the otter in Poland. In Duplaix, N. and Kempf, C. (1985), *loc. cit.*, in press.

Wobeser, G. (1976). Mercury poisoning in a wild mink. *J. Wildl. Dis.*, 12, 335–340.

Wobeser, G., Nielsen, N. O. and Schiefer, B. (1976). Mercury and mink. II. Experimental methylmercury intoxication. *Can. J. Comp. Med.*, 40, 34–45.

Wood, M. (1978–1979). Artificial otter holts. *Water Space*, Winter 1978/9, pp. 16–19.

Wren, C., MacCrimmon, H., Frank, R. and Suda, P. (1980). Total and methyl mercury levels in wild mammals from the Pre-Cambrian Shield area of South Central Ontario, Canada. *Bull. Environ. Contam. Toxicol.*, 25, 100–105.

Yarrell, W. (1841). *A history of British fishes*, 2nd edn. London.

Zaret, T. M. and Rand, A. S. (1971). Competition in tropical stream fishes: support for the competitive exclusion principle. *Ecology*, 52, 336–342.

Index

Printed in the United Kingdom by
Lightning Source UK Ltd., Milton Keynes
139554UK00001B/71/P